I0825149

EMPIRE OF SKULLS

ALSO BY PAUL STOB

Intellectual Populism: Democracy, Inquiry, and the People

William James and the Art of Popular Statement

Thinking Together: Lecturing, Learning, and Difference in the Long Nineteenth Century (ed.)

EMPIRE *of* SKULLS

Phrenology, the Fowler Family, and a New Nation's Quest to Unlock the Secrets of the Mind

PAUL STOB

COUNTERPOINT
CALIFORNIA

EMPIRE OF SKULLS

This is a work of nonfiction. However, some names and identifying details of individuals have been changed to protect their privacy, correspondence has been shortened for clarity, and dialogue has been reconstructed from memory.

First Counterpoint edition: 2026

ISBN: 978-1-64009-683-7

The Library of Congress Cataloging-in-Publication data is available.

Jacket design and illustration by Victoria Maxfield
Jacket image of phrenology bust courtesy of Paul Stob
Book design by Olenka Burgess

COUNTERPOINT
Los Angeles and San Francisco, CA
www.counterpointpress.com

Printed in the United States of America

1 3 5 7 9 10 8 6 4 2

Dedicated to the Misfits, who,
like the Fowlers, want your skull

CONTENTS

EMPIRE OF SKULLS

PROLOGUE

SECRETS OF THE PIRATE'S HEAD

THE SLOOP SET SAIL FROM LONG ISLAND BEARING SOUTH, a crisp night air sweeping along the deck as it went. Soon, the pirate knew, darkness would shroud the vessel, and they'd be far enough from shore that no one would hear the screams. Below deck was a stash of gold he planned to pilfer, but only after he dispatched his shipmates: George Burr, captain of the ship and not yet forty years old; Oliver Watts, twenty-three years old; and Oliver's brother Smith Watts, nineteen years old. The pirate was bigger and stronger than all of them. Plus, he had his axe nearby.[1]

When night fully engulfed them, Smith and Captain Burr retired to the cabin, leaving Oliver at the helm. The pirate, who was using the pseudonym William Johnson, decided to keep him company. After pacing around the creaky deck, he asked Oliver if he could steer. The young man was fine with that; it would give him a break, and they had plenty of time on the voyage.

After some chitchat about life and weather, Johnson squinted toward the horizon and pointed. "Is that Barnegat Light?" he asked, wondering if the lighthouse at the northern tip of Long Beach Island was coming into view.

"No," replied Oliver, explaining that it would be at least an hour before they could see Barnegat.

Squinting harder, Johnson asked again, this time insisting that it must be the lighthouse. Oliver turned to take a closer look—a big mistake.

In a fluid motion, and with a kind of emotionless passion only a veteran killer could manage, the pirate wrapped his fingers around the axe and buried it in the back of Oliver's skull. A crimson geyser twinkled in the moonlight, and Oliver dropped to his knees. Johnson tugged the axe out of his skull and planted it in the young man's back to finish the job.

Just then Smith, awake from the commotion, poked his head out of the cabin door to see what was afoot. Stomping over to him quickly, Johnson swung the axe in a powerful arc, severing Smith's head cleanly from his body. Another fountain of blood sprayed the deck red.

Storming into the cabin, Johnson found Captain Burr sitting up in bed. "What's happening?" Burr asked. When his eyes adjusted to the sight of a large man covered in blood and gripping an axe, he had his answer.

Leaping out of bed, Burr began grappling with the maniac, trying to pry the axe from Johnson's bloody hands. They tussled for several minutes, and Burr proved stronger than his 135-pound frame looked. Maybe getting the booty wouldn't be as easy as the pirate had assumed.

But Johnson was able to push off Burr, creating some distance between them, and swing the axe one more time, lopping off half of Burr's head, one of his eyes, and part of his nose. Blood, brains, and skull fragments decorated the cabin, which now looked less like sleeping quarters than like an abattoir.

When Johnson returned to the deck to dispose of the bodies of the Watts brothers, he found Oliver somehow still alive and shambling zombie-like toward him. Stunned, he held the axe in both hands—one hand at the bottom, the other right below the metal head—and used it to push Oliver over the edge of the ship. Somehow the determined zombie grabbed on to the railing, dangling off the ship's side and refusing to die.

After trying and failing to pry Oliver's fingers loose, the pirate realized there was only one thing left to do. He raised the axe and brought it down on the young man's hand, severing Oliver's fingers and thumb. Oliver fell into the sea as his digits landed on the ship's deck.

Back in the cabin, Johnson helped himself to some beer to take the edge off. Then he grabbed Captain Burr's bloody body, dragged it onto the deck, and pushed it over the side. Next he dragged the headless corpse of Smith and tossed it into the frigid waters.

Wait, the pirate thought, freezing in place. *Where is Smith's head?* He looked around the deck but saw nothing.

Heading back into the cabin, the pirate got on his hands and knees and started feeling around in the darkened corners. Eventually his hand touched some wet, sticky, matted hair. Grabbing hold of Smith's head, he carried it like a small sack of onions onto the deck and, standing in the twinkling starlight, smeared with the blood and brains of his shipmates, launched it into the water, followed by his trusty axe.

The pirate stuffed all the money and valuables he could find into a sack, climbed into the ship's yawl, lowered it into the sea, and rowed away as fast as he could.[2]

IT DIDN'T TAKE long for other sailors to find this ship without a crew. Nor did it take long to discover that the deck was awash in blood. The sailors quickly contacted the authorities, who brought the sloop into port and started searching it for clues. Other than blood and axe marks, the best clue they found was Oliver's severed fingers and thumb.

Determined newspaper reporters got word of the grisly ghost ship and descended on the scene, eager to craft high drama for the daily papers. They worked up colorful columns about the "pools of blood" and "human hair matted with gore" smeared across the ship. The scene had the makings of a crime of the century, and the reporters pledged to embed themselves with police until the killer was caught.[3]

After asking around different Long Island ports, police found the ship's manifest. Immediately one name stood out—William

Johnson. Captain Burr and the Watts brothers were known to everyone in the area, but not Johnson. Plus, people had remembered seeing a burly, gruff-looking man trailing the captain in the days before the voyage.

Then they got the tip that would break the case wide open: A large man—drunk, bloody, and carrying a sack of money—had been spotted in Manhattan. Police raced to the scene only to discover that the man had just left the city with his wife and five-year-old son in tow. They also learned his name was not William Johnson; it was Albert Hicks.

Hicks, meanwhile, took his wife and son to Rhode Island to lay low for a while. They rented a room in a flophouse and waited. After only two days, cops stormed into the flophouse at two in the morning to find Hicks asleep. When he woke up, he knew the jig was up and went peacefully, leaving his wife and son, who were apparently unaware of his crimes, to find their own way back to New York.

The trial of Albert Hicks began on May 18, 1860. Because detectives had never found the bodies of Burr and the Watts brothers, prosecutors worried that they might not be able to secure a conviction for murder. So they went with a charge of piracy, given that the crime had happened at sea and that a piracy conviction carried the death penalty. The trial lasted less than a week, and the jury took fewer than ten minutes to find Hicks guilty. After the verdict was read, the judge sentenced Hicks to death and set the execution date for Friday, July 13—a good day to hang a stoic axe murderer.

That bright summer day became a festival. Over one thousand people rode on the big boat carrying Hicks to Bedloe Island so they could watch his execution. Drinking the saloon on the ship dry, they stumbled over one another to catch a glimpse of the famous criminal and to watch his neck snap at the end of a noose. They weren't the only ones. For only one dollar, curious New Yorkers could secure a spot on a viewing boat, consume unlimited beer and oysters, and witness the death of the man who would become the last pirate of New York.

Shortly after eleven in the morning, the hangman put a black hood over Hicks' head and fastened a noose around his neck. After receiving the signal to proceed, the executioner pulled a lever on the gallows, which dropped a set of heavy weights and pulled Hicks into the air. His neck snapped, and for several minutes, as his spirit left this world, his body twitched in the air, dancing the dance of the condemned. "Few men have passed from earth so wholly unregretted as did this murderer and pirate," wrote *The New York Times*.[4]

THESE GRISLY AXE murders raised some pressing questions for Americans of the time: How could Hicks have brutally slaughtered three innocent men, stolen their money, and then flaunted it around Manhattan without a care in the world? How could a supposedly advanced nation be dealing with such primitive depravity? What made Hicks become a monster?

Hicks himself had answers to these questions. In a lengthy confession given to a reporter prior to his execution, he explained that there were actually five people on the ship that night: Captain Burr, the two Watts brothers, Hicks, and the devil. "I have long felt as though I were the Devil's own," Hicks explained, "and that though he had served me so many years, I must at last be his." The devil craved not just the blood of Hicks's victims but the pirate himself in hell.[5]

The devil's participation in the grisly murders made a good deal of sense to many at the time. Of course the devil was loose on the ship. Hicks's atrocities seemed characteristic of the violence of the 1850s that would shortly open the nation's veins in civil war. The best explanation for Hicks's evil came down to forces outside of him—forces in a supernatural plane of existence that accounted for all that was good and bad in the world.

But maybe there was another explanation. Two weeks before Hicks's execution, authorities locked him in shackles, loaded him into a horse-drawn wagon, and transported him to a business located at 308 Broadway. When they arrived, they parked in front of a window displaying rows of human skulls, which looked blankly ahead as the murderous pirate stepped out of the wagon and proceeded inside, passing under a sign that read "Fowler and Wells, Phrenologists."

Inside, Hicks and the police creaked across wooden floorboards toward the back of the store. As they passed, Hicks saw many, many more skulls watching him. The walls were lined with them—a veritable army of the dead staring at the infamous pirate through hollowed-out sockets. Not all of them were actual skulls, however. Interspersed among them were plaster busts, casts, and masks taken from human heads. Hicks shuffled past the heads of great statesmen, including John Quincy Adams, James K. Polk, Daniel Webster, Henry Clay, and Aaron Burr. There were also the heads of prominent reformers, including William Ellery Channing, Horace Greeley, and Horace Mann. There were the heads of powerful leaders of other nations, including Napoleon Bonaparte and Black Hawk. And there were the heads of evil people just like him—murderers, rapists, and thieves. There was even the head of another pirate, Charles Gibbs, who was convicted of mutiny and murder and hanged in 1831.

After his slow march past the skulls, Hicks met a genial-looking bearded man at the back of the establishment. The bearded man's name was Lorenzo Niles Fowler, and he led the visitors through a door and into an examination room. Stately and grand, the room had high ceilings and intricate crown molding. On the walls were photos, daguerreotypes, and paintings of notable people the world over. Lining the room were two writing desks and a plush Victorian couch. That was where ladies and gentlemen could watch their friends and family receive an examination. In the middle of the room was the exam chair, where Hicks sat down, his shackles clanging against the wood.

Lorenzo gauged Hicks's head visually, getting an overall impression of its organization and size. Then his fingers began gliding expertly across the pirate's cranium. He felt for bumps, depressions, masses, and spans, moving from the base of the skull to the top of the forehead, along the sides, and even around the eye sockets. As he worked, he called out different numbers to his assistant, who recorded Lorenzo's findings.

"Four," Lorenzo announced after touching one part of the head. "Six," after another. "Seven," if the area was large, and "One," if the area was small.

Hicks's examination took only a few minutes, but it produced exactly what the police were hoping for—a scientific account of Hicks's actions. "Phrenological Character of Albert W. Hicks, Given at Fowler and Wells's Phrenological Cabinet" was the title of the report. It appeared in a pamphlet alongside the pirate's confession, trial, and biography, available on newsstands the morning of his execution.[6]

Lorenzo's report noted the power of Hicks's physical attributes, including his mental faculties, but lamented that those faculties were "susceptible of intense feeling" and the "heated impulse of passion." His aptitude for intense feeling was combined with a large brain capable of "a great amount of general mental power." On top of that was Hicks's knack for working with his hands. With "a good degree of order and arrangement" and "native talents for making estimates and calculations," he would thrive in "any kind of mental operation where order, method, system, knowledge of principles and places is required." He could have been a carpenter or brickmason, maybe even an engineer or mechanic. Too bad he was never part of a structured environment where his natural talents could help him become a contributing member of society.

But it got worse. Without proper development, Hicks's natural aptitude for hard work, arrangement, and calculation dwindled just as his "Combativeness" grew stronger. Lorenzo's report noted that the part of Hicks's brain responsible for pugnacity was far too developed. Combined with a strong will and spirit of resistance, Hicks was prone to lash out at the world, especially because his "love of property," also known as Acquisitiveness, pushed him toward "the gratification of his various desires." These overly developed mental faculties, Lorenzo lamented, stood in stark contrast to his underdeveloped Spirituality and Veneration—faculties responsible for religion and morality—leading Hicks "to act without due regard to the Higher Power, and without feeling his dependence on, or much responsibility to, his Creator."

All these details led Lorenzo to a striking conclusion:

> The crimes that he has been led to commit are full as much the result of a want of the right kind of education, as from his natural organization. He has strong passions, and an unbending and headstrong will; but with proper culture, and good circumstances, he would, most likely, have used his energy and talents in a way to secure success and respectability, instead of warring upon the rights and interests of his fellow men.

Here, then, was a radically different account of the convicted pirate's actions. His brutal murders were not due to the devil's influence, nor were they the result of his nature; Albert Hicks was not born an axe murderer. His actions stemmed from a mental makeup that was not properly cultivated, as he lacked the education and environment needed to carve out a morally upright position in society. In that way, Hicks was only partially responsible for his crimes. His family, friends, and community also bore some responsibility. America too. The nation's hands were spattered with the blood of Burr and the Watts brothers right along with the pirate's hands.

That's why America had to do better. To avoid another Albert Hicks, it had to do a better job educating children. To make brutal murders a thing of the past, it had to help suffering people and prop up broken families. To live up to its ideals, the fledgling democratic country needed to break the shackles that held people back. To do these things, America could and should pray to God. But equally as important was understanding human nature, which required one thing above all: phrenology.

DISCOVERED IN THE late eighteenth century by a German physician named Franz Joseph Gall and propagated around Europe in the early nineteenth century by Gall and his assistant, Johann Spurzheim, phrenology trickled onto U.S. shores in the 1820s and took root in the 1830s. A science of character and human nature, it purported to pinpoint a person's inner being—specifically his or her mental makeup—via the shape, contours, masses, and spans of his or her head. Meaning "discourse on the mind," phrenology was built on two key ideas. First, the shape of a person's skull corresponded with the shape of that person's brain, meaning that a bump or cavity on the skull mirrored a bump or cavity on the brain. Second, the brain was not a unified mental mass but composed of individual organs that performed different functions. Phrenology was the study of those individual organs and the ways they combined to create distinct character traits.[7]

Although phrenologists did not fully agree on the number of organs in a person's brain, the general consensus settled on thirty-seven. The organs ranged from "Combativeness" and "Destructiveness" to "Amativeness" (sex drive), "Philoprogenitiveness" (love of children), and "Inhabitiveness" (love of home). There were also organs related to reason and intellect, including "Causality," "Comparison," "Eventuality," and "Mirthfulness" (humor). And there were organs related to perceptions of the external world, including "Form," "Size," "Weight," "Time," "Tune," and "Language." Organs corresponding to man's higher nature and spiritual existence included "Veneration," "Benevolence," and "Hope."

In Europe, phrenology was primarily a science vying for status among other "natural philosophies" that supposedly defined humanity. In the United States, phrenology remained a science, but it also assumed important practical dimensions. Practical phrenology, as it was known, was an effort not just to gauge people's organs and to define their character but to give them tools for enhancing their mind and progressing as human beings. American phrenologists maintained that people could literally, through concerted effort, grow their mental organs. Thinking in particular directions and

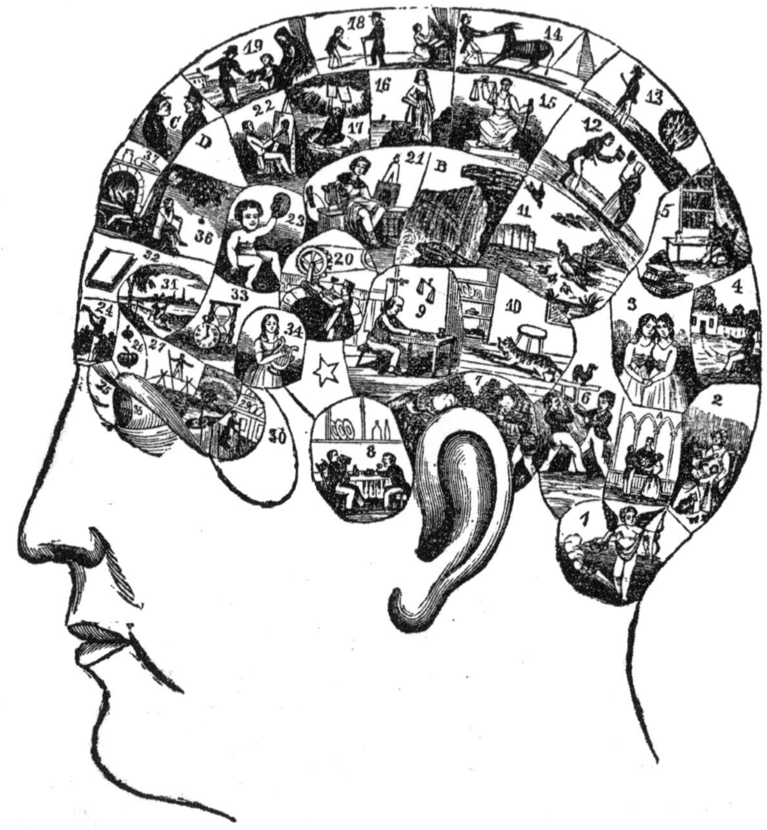

NUMBERING AND DEFINITION OF THE ORGANS.

1. Amativeness, Sexual and connubial love.
2. Philoprogenitiveness, Parental love.
3. Adhesiveness, Friendship—sociability.
A. Union for Life, Love of one only.
4. Inhabitiveness, Love of home.
5. Continuity, One thing at a time.
6. Combativeness, Resistance—defence.
7. Destructiveness, Executiveness-force.
8. Alimentiveness, Appetite, hunger.
9. Acquisitiveness, Accumulation.
10. Secretiveness, Policy—management.
11. Cautiousness, Prudence, provision.
12. Approbativeness, Ambition—display.
13. Self-Esteem, Self-respect—dignity.
14. Firmness, Decision—perseverance.
15. Conscientiousness, Justice—equity.
16. Hope, Expectation—enterprise.
17. Spirituality, Intuition--spiritual revery.
18. Veneration, Devotion—respect.
19. Benevolence, Kindness—goodness.
20. Constructivness, Mechanical ingenuity.
21. Ideality, Refinement—taste—purity.
B. Sublimity, Love of grandeur.
22. Imitation, Copying—patterning.
23. Mirthfulness, Jocoseness—wit—fun.
24. Individuality, Observation.
25. Form, Recollection of shape.
26. Size, Measuring by the eye.
27. Weight, Balancing—climbing.
28. Color, Judgment of colors.
29. Order, Method—system—arrangement.
30. Calculation, Mental arithmetic.
31. Locality, Recollection of places.
32. Eventuality, Memory of facts.
33. Time, Cognizance of duration.
34. Tune, Music—melody by ear.
35. Language, Expression of ideas.
36. Causality, Applying causes to effects.
37. Comparison, inductive reasoning.
C. Human Nature, perception of motives.
D. Agreeableness, Pleasantness—suavity

The "symbolical head," a widely reproduced image of scenes portraying phrenological organs and their location on the head

refraining from thought in other directions could change the size of different parts of the brain. By reading more, practicing kindness, socializing with friends, working on math problems, praying and worshipping more, and so forth, you could actually become a different person.

This emphasis on change gave practical phrenologists their most important idea, one that attracted millions to their cause: improvement. Because you had the power to exercise, or refrain from exercising, your organs, you also had the power to improve yourself and to help others become better people. Phrenological examinations in the United States usually included specific directions on how to improve your life—from career advice or the kind of person to marry to the management of the home, which activities to pursue, and more. In many ways, practical phrenology was an early iteration of what would become known later in the nineteenth century as the "self-help" movement. But phrenology's goals were loftier than that name suggests. Self-help implies individual improvement, which was key to phrenology as well. But phrenologists also insisted that their science was the secret to improving families, communities, societies, and nations. Ultimately, phrenologists in the United States saw their work as the scientific underpinnings needed for a new nation to create heaven on earth.

As it turned out, this was exactly the message that Americans at the time wanted and needed to hear. Beginning around 1830, the United States entered what historians have dubbed the "era of reform," a multifaceted, decades-long push to fix the problems of the nation that survived its founding. Painfully aware of the country's sins, reformers got together and started movements to end slavery, to give women the right to vote, to curb the evils of alcohol, to eat only vegetables, to improve the care of prisoners and mental-health patients, to help the poor, to better educate children, to worship God in new and exciting ways, and more. The pulsating energy for change prepared the way for a new science of the human mind that sought widespread progress for the nation. Little wonder that many prominent reformers embraced phrenology—abolitionists William

Lloyd Garrison and John Brown, suffragists Elizabeth Cady Stanton and Susan B. Anthony, intellectuals Henry Ward Beecher and Ralph Waldo Emerson, writers Horace Greeley, Edgar Allen Poe, and Walt Whitman, among many others. Because practical phrenology maintained that everyone could improve themselves and their communities, no matter where they started in life, reformers saw it as the ultimate scientific tool for lifting the downtrodden to a higher place in American society.[8]

But it wasn't just notable reformers who embraced phrenology in the decades leading up to the Civil War; the science captivated ordinary Americans too. In cities and towns across the country, people flocked to lecture halls to hear about its wonders. They paid traveling phrenologists for a scientific reading of their character and advice on how to live. They bought books and periodicals for home study, which was increasingly prevalent at a time when literacy rates were rising and the cost of printed material was falling. They acquired skulls, busts, and casts of human heads to better understand what the science could do for their world. From the 1830s until at least the Civil War, and to a lesser degree in the decades beyond, phrenology flourished as a cultural and scientific phenomenon unlike America had ever seen. It was the right science for the right time—a time when an adolescent nation needed grounding, direction, and something to believe in. And it was thanks ultimately to the efforts of one family.[9]

THE FOWLERS WERE the preeminent proponents of practical phrenology. Their last name was inseparable from the science, and they were either celebrated or reviled, depending on one's point of view, for spreading the mental philosophy to all corners of society. Millions of Americans knew them, and not just by reputation; many had their head examined by one of the family members or knew someone who sat for a phrenological reading and revelation of mental prowess.

The Fowlers were led by three siblings, all of whom were

Orson Squire Fowler (1809–1887)

responsible for making practical phrenology a cultural phenomenon. The eldest sibling, Orson, was born in 1809 and the first to embrace the science. A serious young man and even more serious adult, Orson—or "O. S." as he was known to family and friends—sported medium-length hair and a bushy beard that made him appear much older. Looking at the world through intense eyes and a tall forehead that signaled his intellect, he was a lofty dreamer, tireless worker, and shrewd entrepreneur who was convinced early on that phrenology was not just a new take on the human mind but something that could transform the world. Treating the science as gospel, he spread a message of salvation to all corners of the country and baptized multitudes into his church of progress.

Middle sibling Lorenzo, born in 1811, was the most scientifically astute of the family. More personable than his older brother, Lorenzo—or "L. N."—had kind eyes, an easy smile, a large forehead, and a robust constitution. Like his brother, he sported

Lorenzo Niles Fowler (1811–1896)

medium-length hair and a bushy beard that aged him. For both Fowler brothers, who started their advocacy in the 1830s when they were in their early twenties, appearing older gave them a modicum of credibility as serious scientists.

Third sibling Charlotte was born in 1814 and instrumental to the family's success. She worked behind the scenes to balance the books, to grow the business, and to ensure that the lectures, writings, and examinations of her brothers happened without issue. Pretty, delicate, and kind, she also appeared matronly when still a young woman. That persona served her well in the long run; after her brothers left the family business around the time of the Civil War, Charlotte took control of the entire operation and had a long, distinguished career as a publisher and reformer.

Also crucial to the growth and success of practical phrenology in the United States were the spouses of the Fowler siblings. Orson's wife, Eliza, whom he married in 1835, supported his wild scientific

Charlotte Fowler (Wells) (1814–1901)

dreams, though she remained largely in the background. Lorenzo's wife, Lydia, whom he married in 1844, was a scientific pioneer in her own right. Studying and promoting phrenology, especially for women and children, she became the second woman to earn a medical degree in the United States and the first female professor of medicine in the country's history. She and Lorenzo formed the first power couple of phrenology. The other power couple was Charlotte and her husband, Samuel Wells, whom she married in 1845. Together, Charlotte and Samuel created one of the most influential American publishing firms of the mid-nineteenth century.

Orson, Lorenzo, and Charlotte, along with Lydia and Samuel, worked shrewdly to make phrenology central to the nation's consciousness. They did so by delivering thousands of lectures in America's biggest cities and smallest towns; by examining the heads of an untold number of people and giving them directions on how to create a better version of themselves; by publishing wildly popular books on a host of topics—phrenological and otherwise—and creating one of the most influential periodicals of the day, *The American Phrenological Journal*; by connecting their ideas with the work of countless other reformers, thereby positioning phrenology as the scientific basis for social change; by showing how their work could explain everything from murder, rape, and insanity to wealth, power, and celebrity or race, gender, God, and more. Along the way, newspapers heralded their work for astonishing "multitudes by the almost unerring accuracy" of their readings; for "exciting inquiry and securing converts" to their science; and for becoming "the best known and most popular speakers and writers upon the human system" anywhere on the planet. "Go WHEREVER the English language is spoken," Orson boasted in the early 1870s, after having championed phrenology for almost forty years, "and you will not talk long among intelligent persons about O. S. and L. N. Fowler, without hearing some one of them tell of some remarkable phrenological hit made by one of these phrenological brothers on some ancestor or acquaintance of somebody."[10]

Beyond championing phrenology's message of improvement, the Fowlers used the science adeptly to navigate an age-old question that seems to captivate every generation: What makes us who we are, nature or nurture? The question had been debated since ancient Greece and became especially pressing in the Enlightenment amid the rise of empiricism as thinkers grappled over whether our inner being or our environment played the largest role in our development. For many religious believers, the answer was that God created us to be who we are (nature). For others, especially as the era of observational science spread across the Western world, the answer was that people become who they are through their upbringing (nurture). The Fowlers came along and offered practical phrenology as proof that both nature and nurture shape us and are ultimately inseparable. People have natural endowments; they're born with a particular set of mental faculties. But everyone can change those faculties through hard work, turning their faculties into something new, something better, something stronger. As Orson declared in his magnum opus *Human Science*, a summary of his life's work published in 1870, "Nature rewards its improvement with the richest bounty possible to receive." With this answer to the age-old debate, Americans had a clear, hopeful, scientifically structured path forward.[11]

Another reason millions of Americans embraced practical phrenology was because the Fowlers tapped into the widespread, long-standing impulse to measure ourselves scientifically. When ordinary Americans visited the Fowlers for a phrenological reading, they came away with a chart that listed their thirty-seven phrenological organs, each one assigned a number ranging from one to seven. Taken together, the numbers constituted a trustworthy map of character, a numerical encapsulation of what lay inside them, which was inaccessible at the time to any medical procedure or psychological probing. The only true way to get inside us, the Fowlers maintained, was through their science. As a result, the piece of paper generated from a phrenology exam often became a prized possession, handed down to children at the end of life. The chart defined people not as a haphazard stream of experience or a ceaseless swing of emotions but as a specific, quantified, scientifically delineated mind, which could grow and improve through concerted effort.[12]

In point of fact, we do much the same thing today, hoping to measure ourselves scientifically and better know who we are. We take IQ tests to quantify our intelligence, Myers-Briggs and Enneagram tests to categorize our characters, and at-home DNA tests to chart the very fabric of our being. Indeed, when you spit into a tube and mail it to a testing company, you will receive online access to a chart that details your genetic heritage, your health tendencies, your aptitudes, and more. Numbers, charts, and graphs about ourselves, which we pay for willingly and gladly, provide us a measure of identity that points the way forward. The impulse that led millions of Americans to embrace practical phrenology is, ultimately, the same impulse that drives us to acquire data about ourselves today.

DESPITE THE FOWLERS' incredible success and widespread social, cultural, and scientific influence in the nineteenth century, few people remember them today, the result of the ways we tend to talk about phrenology. In general, phrenology is considered either a punchline about foolish theories of the past or a racist pseudoscience that occupies a lamentable place in our collective history. Both ways of talking about phrenology have validity, but they ultimately gloss over a rich, strange, complicated history that echoes around us to this day.

As a punchline, phrenology appears regularly in popular culture. You've no doubt seen remnants of the Fowler empire in the form of an image depicting a person's head in profile with different areas marked off and labeled. These areas and labels, which people often change to humorous effect, supposedly represent the mind underneath the head in profile. This image is a play on a famous engraving that the Fowlers created and used in numerous publications. Then there's the phrenological bust, a plaster or porcelain head also with different areas marked off and labeled to show the location of the various phrenological organs. The bust, which was created and popularized in the United States by Lorenzo Fowler, appears regularly in movies, TV shows, cartoons, and scenes designed to portray vaguely medical yet ultimately comical settings.

Case in point, an old episode of *The Simpsons* (season 7, episode 8, to be precise) includes a wonderful phrenology gag. Sitting in his cavernous office at the nuclear plant, Mr. Burns talks with a couple of FBI agents, who appear to be straight from *Dragnet*, about a woman he recently saw in the post office. Thirty years prior, Burns believes, this woman (who happens to be Homer Simpson's long-lost mother) destroyed some of his germ-warfare research.

One of the FBI agents holds up an old wanted poster and asks, "Are you sure this is the woman you saw in the post office?"

"Absolutely!" Burns exclaims. Without missing a beat, he lifts a phrenology bust from under his desk and measures it with an old set of calipers. "Who could forget such a monstrous visage? She has the sloping brow and cranial bumpage of the career criminal."

Waylon Smithers, Burns's trusty assistant, then chimes in sheepishly: "Ah, sir, phrenology was dismissed as quackery 160 years ago."

"Of course you'd say that," Burns retorts, now using the calipers on Smithers's head. "You have the brain pan of a stage-coach tilter!"

This scene is part of a running joke about how old Mr. Burns really is. It's also about our collective scientific gullibility. Science of the past looks silly today, allowing us to wonder smugly how people could have believed such foolishness. They were either dumb or duped, taken in by hucksters who wanted to empty their pockets with dreams of cranial bumps and personal improvement.

But there's a more troubling way we talk about phrenology today. Even a quick search online will reveal that the Fowlers' science is the poster child of scientific racism. With phrenology, reports one article, "promoters of anti-Black racism and white supremacy . . . co-opted the authority of science to justify racial inequality." According to another article, phrenology is the story of "racial characteristics used to motivate policy and cultural shifts that privileged certain whites and oppressed or marginalized peoples of color and other whites." The ill-gotten goods of these scientific racists, we are told, are now preserved in museums and universities, a shameful part of our past. There's even an article on the Fowlers specifically with the provocative title, "How Profit and Prejudice Built a Family's Human Skull Collection." Then there's a story in *The New York Times* about how the Fowlers' "racially tinged" history prompted one Manhattan restaurant to change its name from Fowler and Wells to Temple Court. The original name came from the fact that the restaurant is located on the site of the Fowlers' old Phrenological Cabinet; the new name is designed to distance the restaurant from promoters of racial division.[13]

There is no question that phrenology has a problematic past when it comes to race. Some of its supporters were slaveholders and bigots who used the science to justify the subjugation of people who weren't white, formulating a racial hierarchy that placed Caucasians at the top, duty-bound to rule over other races for the sake of humanity. But to peg phrenology as *only* a racist pseudoscience misses the point of an intellectual movement aimed squarely and earnestly at progress, including for people of color. No doubt the movement fell short of its own ideals. No doubt phrenology's proponents said, wrote, and did things that were abhorrent when it came to other human beings. No doubt the racial imagination of many phrenologists was harmful, and we certainly shouldn't look to phrenology today for any insights on race, progress, or justice.

But the story of phrenology isn't just a story of racial subjection; it's more complicated than that, especially when it comes to the Fowlers. At times both Orson and Lorenzo issued painfully sweeping generalizations about different races (for example, "African skulls are usually thicker than Caucasian" due to lesser brain activity), and they reverted to old racist tropes (for example, Jews are "remarkable first for their love of money"). Yet they adamantly opposed slavery, affirmed the scientific value of race mixing, and argued for the unity of the races at a time when other scientists believed that races had different origins. These contradictions were quite common at the time, and certainly not limited to the Fowlers, which is not to diminish the real problems of their writings. But it's crucial to understand their ideas in the context of pre–Civil War America and in the matrix of other conceptions of the world then swirling around.[14]

The Fowlers and their supporters—white and Black—embraced phrenology specifically *because* it promoted equality and liberation and was a step ahead of other scientific frameworks of the day. Because everyone could improve himself or herself, no matter where he or she started in life, everyone could become a better version of himself or herself. There was real hope in that view, which was why the Fowlers drew together reformers who saw phrenology's potential to help liberate enslaved people, to advance the place of women in society, to educate all Americans, to abolish the death penalty, to curb the evils of alcohol, to make prisons more humane, and much more. Simply put, applying the label "racist pseudoscience" to phrenology encourages us to gloss over a complicated movement that was ultimately more progressive, even radical, than we typically consider.

Even using the word *pseudoscience* to discuss phrenology, as often happens, is misleading and ultimately unhelpful. Today, we know that phrenology is not a scientifically accurate characterization of the brain and the mind; neuroscientists at the University of Oxford conclusively proved as much in 2018 using MRI scans. Even in the nineteenth century, many scientists understood the shortcomings with the theory. What's more, the term *pseudoscience* was actually coined as a characterization of phrenology. In 1859, the famous physician and man of letters, Oliver Wendell Holmes Sr., sat for a phrenological reading and heard nothing but flattery of his character, which, he concluded, was a big reason people supported it. In an article in *The Atlantic* prompted by this experience, Holmes proposed the term *Pseudo-science* as a nomenclature for "a self-adjusting arrangement, by which all positive evidence, or such as favors its doctrines, is admitted, and all negative evidence, or such as tells against it, is excluded." According to Holmes, phrenologists accepted whatever "data" supported their work and dismissed the mountain of data that proved them wrong.[15]

Despite Holmes's insights, and despite what we know today about how the brain works, calling phrenology a pseudoscience doesn't get us closer to understanding its place in the dynamic world of nineteenth-century America. Proponents of a particular science, especially a new one, refer to themselves as scientists, not pseudoscientists; the term *pseudoscience* is used only by people who oppose a particular intellectual movement. At the same time, the label obscures the fact that science is always contested, always in the process of refinement, always undergoing revision. In the pre–Civil War decades, phrenology was a legitimate possibility as an accurate science of the mind. There were problems with it, to be sure, as there are with every science (even today); but supporting it was neither foolish nor faddish. In fact, many of the scientists who objected to it at the time did so because their own understanding of the brain and the mind was wrong. For much of human history, and well into the nineteenth century, the brain was considered a singular organ that carried out one function, then another, then another. It was the phrenologists who popularized the theory that the brain comprises multiple organs responsible for doing different things and doing them at the same time. Today, this theory is known as cortical localization, and it remains central to modern neuroscience. Phrenology was an important step forward for our understanding of the brain and the mind, despite the fact that it was ultimately inaccurate.[16]

THE COMPLICATED STORY of phrenology in America is the story of how the Fowlers popularized a new science, got millions of people excited about it, and built an empire in the process. Indeed, Orson, Lorenzo, and Charlotte, along with their spouses and associates, gave the American people answers about their future at a time when anxious citizens of a fledgling nation desperately needed answers in the form of perhaps the most representative symbol of humanity—the skull. Their message was simple: The secrets of human life could be found in something everyone already had. What was needed was a proper method for getting at it. When they got to what was inside of them, people could turn their personal insights into knowledge of the world around them and purposeful, meaningful progress.

In sharing this message, the Fowlers got their hands on lots of skulls—actual skulls that had once been inside living, breathing human beings. And the more they amassed, the more they opened a new reality to the American people. At the Phrenological Cabinet in New York, the wildly popular tourist destination the siblings created, anyone who stepped through the doors could hold the Fowlers' skulls and touch the truth of humanity. Feeling the heads of pirates and presidents, savants and savages, Americans could grasp reality in their hands. And for a nominal fee, they could learn the secrets of their own inner being in the examination room at the back of the Cabinet.

With each person who came through the door, with each book they sold, with each lecture they delivered, with each examination they performed, the Fowlers built a business, then a movement, then an empire. Atop a mountain of skulls, they led the adolescent nation through some of the most transformative decades in its history.

CHAPTER ONE

A MEETING OF MINDS

IT WAS A RATHER DEVIOUS PLAN THE STUDENTS HAD CONcocted. They would host a mock debate where both sides would prove the same thing—that this new so-called science of phrenology was wrong and foolish. It had been making news across the country recently, but they, members of Amherst College's Society for Natural History, class of 1834, were no rubes easily taken in by absurd fads. There was more: Not only would the debate expose the farce of phrenology for the entire student body, but the students would have a little fun at the expense of that "tall, grave, sober fellow" who seemed taken by the science. Fowler was his name.[1]

The question at the center of the Amherst debate was "Is phrenology entitled to the name of science?" One student, Alonzo Gray, agreed to argue the affirmative side. For the negative side, the Society for Natural History had an ace up its sleeve—a student known for his oratorical prowess and incisive wit. The son of the most famous minister in the country, and on his way to becoming perhaps the most famous man in America, he was sure to trounce all over the science and anyone who supported it. His name was Henry Ward Beecher.[2]

Talented orator that he was, Beecher knew he had to understand phrenology if he was going to tear it apart. So he contacted a publisher in Boston—Marsh, Capen, and Lyon—and acquired a stack of books, immersing himself in silly ideas about heads and bumps.

Then something strange happened. The more Beecher read of phrenology, the more it intrigued him. The more he studied the strange new science, the more sense it made.

When it was time to debate, Beecher delivered a brilliant speech, as he was wont to do, but it wasn't against phrenology; it was in *support* of the new science. To the shock of everyone in attendance—except, perhaps, Orson Fowler—he defended phrenology's basic principles and announced himself a convert. Phrenology may not be fully worked out yet, Beecher maintained, but that day will come.[3]

After the debate, Beecher cornered Orson and asked if he wanted to borrow his stack of phrenology books. The answer was yes, of course, and the two young men became fast friends. In the days, weeks, and months after the debate, they studied phrenology together, practiced on a model head Beecher had purchased, and even gave phrenological examinations to fellow students. They got so good at these exams that they started charging money for their services. For only two cents, Amherst students could learn the secrets of their character, aptitude, and prospects. One student was found to have very large Locality, large Individuality, and large Form, Size, Constructiveness, and Imitation. "This young gentleman," Orson later recalled, "was distinguished for his geographical knowledge, having drawn and published several maps." Other students, who had reputations for egotism and conceit, were found to have Self-Esteem developed "in such a degree as to elongate the head." Even Beecher, known as a slob who threw his clothes and books across his room "in all directions and in utter confusion," was found to have Order "almost wholly wanting."[4]

With the examinations going so well on campus, Orson and Beecher decided to team up for lectures and demonstrations in towns neighboring Amherst, including Enfield, Belchertown, and Ware. A kind of dynamic phrenological duo, Beecher orated about the science while Orson examined the heads of audience members—for a fee, of course. Apparently, in one lecture alone, the pair received a total of ten dollars, a princely sum for two young college students. With his cut, Beecher spent eighty-five cents on an engagement ring for his girlfriend.[5]

Orson and Beecher must have been a sight to behold. Though young at the time, both men would have long, influential public careers, shaping American culture through their speaking, writing, and advocacy. Both would also be embroiled in very public sex scandals later in life. While Beecher never became the kind of practical phrenologist that Orson became, the famous preacher remained forever grateful to the new science for unlocking the secrets of humanity. "All my life long," Beecher recalled in the twilight of his years, "I have been in the habit of using phrenology as that which solves the practical phenomena of life. I regard it as far more useful, practical, and sensible than any other system of mental philosophy which has yet been evolved." Orson believed the same thing, and he made it his life's work to convince the American people that practical phrenology was the scientific tool they needed to improve themselves, their families, their communities, and their nation, regardless of where they started life.[6]

But that was a long way off. In 1832, the promise of phrenology was only partially realized, especially in the United States. Just thirty years old at the time, the science had started across the mighty Atlantic Ocean with a young German man who looked at his classmates and noticed something peculiar.

There's another kid with those bulging eyes. Eyes like saucers. Cow eyes. And this kid also has a great memory. Why can kids like this memorize long passages so easily? A connection between their wide eyes and their memory? Perhaps.

Franz Joseph Gall never recorded exactly what went through his mind when he first noticed the correlation between kids with saucer eyes and strong memories in grammar school. But their physical features and corresponding mental abilities made a deep impression on him, especially when he noticed them again when he was in medical school. "If memory manifests itself by an external character," as Gall later recalled his early thinking, "why should not the other faculties have their characters outwardly visible?"[7]

Perhaps part of his inspiration came from his own outward appearance. Gall was not an attractive man. Balding from an early

Franz Joseph Gall (1758–1828), discoverer of phrenology

age, with heavy, pudgy eyes and a small, weak chin, he was a brilliant thinker without much charisma. Fortunately, the science he discovered wedded intelligence to cranial size instead of physical attractiveness. His high, round, robust forehead signaled a man with brains.

Gall was born in 1758 in Germany, and he loved nature from an early age. As a kid he often ventured into the woods, caught animals, and brought them home as pets, which he studied carefully to understand their behavior. Teenage Gall continued his observations at the University of Strasbourg, where he studied the natural world and its many animals, including human beings. The more he progressed through university, the more he felt called to a career in medicine, and in 1781 he moved to the University of Vienna to pursue a medical degree. Particularly excited about the university's plan to open a new hospital, he couldn't wait to get his hands on patients, including those in the adjoining mental asylum. When the hospital finally opened in 1784, Gall spent countless hours examining the sick and the broken.[8]

Graduating with a medical degree in 1785, Gall opted to remain in Vienna, then the center of the Holy Roman Empire, to practice medicine. Sometime over the next several years, although it's unclear exactly when, he began to develop his new scientific theory in earnest. A pivotal moment came when he met a five-year-old music prodigy named Bianchi.

Bianchi's parents brought the little girl to Gall so he could help them understand her impressive abilities. Apparently, the five-year-old was able to repeat on the piano any piece of music she heard only once or twice. Even after hearing an entire concerto a couple times, she could play it back flawlessly. The parents were stunned, but they noticed something else: Her amazing ability to recall what she heard applied only to music. When it came to other things—verbal material, for instance—Bianchi had an ordinary memory at best.

Examining Bianchi, Gall thought back to the kids with saucer eyes. Their eyes correlated with strong memory for verbal material. But here was little Bianchi, with no saucer eyes and an incredible memory for music. Could it be that memory is not one faculty but many? That is, could different parts of the brain be responsible for different kinds of memory—one part for verbal material, another for musical material, and maybe others besides? The only way to find out was to collect more data.

That meant observing and examining lots of people. In effect, Gall needed heads—as many as he could get his hands on if his research were to have any scientific validity. So he became a kind of headhunter, at least of the scholarly variety. One way he hunted heads was through his fingers. While he could see the bulging eyes of kids with strong memories for verbal material, he needed to touch the lumps hidden under hats and hair that corresponded with other mental abilities. Playing up the role of kindly physician and staid researcher, he convinced people to let him feel, massage, and gauge their head by running his fingers across their cranium. Taking careful notes on what he felt and what people reported about their character, he continued theorizing about inner nature and outer forms.

In truth, the idea of correlating outward appearance with interior mental abilities had been around for millennia. In ancient Greece, such thinkers as Pythagoras, Hippocrates, and Aristotle proposed a correspondence between appearance and the mind, between what people looked like and how they acted. In the Middle Ages, naturalists pushed the point further by maintaining that people with animal-like characteristics often resembled the animals they mimicked. People with lion-like qualities (courageous leadership, for instance) tended to look like lions, while people with donkey-like dispositions tended to look like donkeys. Or so the theory went.[9]

The attempt to correlate appearance and character went further in the eighteenth century via the science of *physiognomy*, a term derived from Greek and meaning "nature" and "to know." Physiognomy maintained that it was possible to know a person's nature by the shape, angles, and features of the face. Beginning around 1770, Dutch physician Petrus Camper argued that facial angles corresponded with intelligence. Basically, if you looked at a person in profile, you could map a facial angle that indicated how smart and refined the person was. Draw a right angle connecting the nostril, ear, and top of the head, then draw a straight line from the nose to the forehead. People with large, protruding foreheads have large facial angles and high intelligence, but the lower the facial angle, the lower the intelligence. A person with a facial angle of 80 degrees was brilliant and highly civilized, while a person with a facial angle of 70 degrees was not as smart or as civilized. Unsurprisingly, Camper maintained that white Europeans had facial angles around 80 degrees, while Asians and Africans typically had facial angles of 70 degrees.[10]

Joining Camper in the promotion of physiognomy during the eighteenth century, and effectively paving the way for Gall, was the Swiss clergyman Johann Caspar Lavater. Lavater's work in physiognomy proved more popular and wide-reaching than Camper's, largely because he wrote as a man of the cloth, which put his ideas before broad audiences. He was also keen on communicating science through imagery. His books on physiognomy contained upward of eight hundred illustrations detailing the connection between outward appearance and inner nature. These illustrations equipped people to identify and classify individuals they encountered, knowing whom to avoid and whom to engage, who was shifty and who was trustworthy, all by appearance alone (an idea that was basically just a shortcut to prejudice).[11]

Gall was thoroughly familiar with physiognomy and the work of Camper and Lavater when he began feeling people's heads. But he wanted to get at something that the physiognomists did not pursue—namely, the role of the brain in the nexus of appearance and the mind. Gall surmised early on that particular physical traits—bulging eyes, for instance—correlated directly with the shape of the brain below. Perhaps parts of the brain pushed on particular parts of the head, resulting in lumps and masses. What's more, perhaps those different parts of the brain performed different functions—bulging eyes and verbal memory formed just one connection. Perhaps mapping bumps on the head with character traits could reveal what different regions of the brain did.[12]

To support this idea, Gall needed more heads. But where could he get them? After exhausting his supply of family, friends, and acquaintances, he started hosting gatherings for people who could use some extra cash—errand boys, servants, coachmen, and more. To smooth over the awkward part where he needed to feel their heads, he plied his guests with beer, wine, and cash. Once they were properly soused, Gall questioned them about their temperament, character, and inclinations. Who among them was quarrelsome? Who was quick to anger? Who was conflict averse? Who wanted to sit back and simply watch the evening unfold? After his guests divulged this information, he examined their heads and found some telling patterns. The heads of people quick to anger, he discovered, were broad and robust right behind the top of the ears—at least compared to the heads of guests who were generally calm and relaxed.

An interesting pattern. But Gall well knew he couldn't build a new science from drunk delivery boys. Playing the doctor card helped him gain access to new heads, as schools, hospitals, and prisons proved relatively amenable to giving a physician access to people in their care. Gall examined hundreds upon hundreds of heads in prisons and asylums, which provided documentary evidence about why those people were in their care. Gall could read their charts and compare their behavior with their cranium.

In one instance Gall compared people in different prisons who were incarcerated for compulsive stealing. There was a young man of only fifteen in Vienna who could not help himself from stealing, "notwithstanding the severest punishments" he received. There was a boy of only twelve in a Bern prison who "could never prevent himself from stealing; with his own pockets full of bread, he still took that of others." Same was true of a convict in Haina, who "stole everything he saw"; "no corporal punishment could correct" his behavior, Gall noted. But he observed an important fact: Not only were they all compulsive thieves but their heads were shockingly similar. They had a small cranium overall, with a "low and depressed" forehead yet large "anterior lateral parts of the temples." Theirs were the heads of people with "mental imbecility" that "excludes all moral liberty."[13]

It was all important grist for the scientific mill. And the more grist he gathered, the more confident he became in fomenting an intellectual revolution.

THIS REVOLUTION WAS based on Gall's answers to two big questions: Where is the mind located? And what does the brain do? Despite the ultimate inaccuracy of his theory, his answers to these questions pushed modern science forward at a pivotal time in world history.

The ancient Greeks held various ideas about the mind. Early on, there was a widely held belief that the heart was the control center of the body, as it was the home of the soul, while the brain played a lesser role in directing human conduct. The mind, therefore, was somewhere around the heart-soul intersection. Other thinkers argued that the stomach was the seat of the mind, while still others, including Hippocrates, identified the brain (although his view was the minority position). Regardless of where they located the mind, they agreed on its three basic functions—sensation (or feeling), cognition (or thinking), and memory (or recalling). This tripartite division was more or less an accepted fact for two thousand years. It was still the prevailing theory at the time Gall began his research.[14]

The ancient Romans had a slightly different view of the mind and body because of a prominent physician-philosopher named Galen of Pergamon. Galen was interested in how bodies interacted with minds—or how material substances could produce immaterial phenomena, like thoughts. To explain body-mind interaction, Galen posited the idea of "spirits" moving through veins. Our liver, he taught, produced "natural spirits," which then traveled to our heart, which converted the natural spirits to "vital spirits," which then traveled up to the brain, where blood vessels at the base of the brain turned them into "animal spirits." These animal spirits were the spirits of the mind. They lived in the ventricles of the brain until the brain sent them to do a job—for instance, moving muscles so a person could walk. When assigned a job, the animal spirits entered the nerves, traveled down the body to carry out the task, then returned sensory information to the brain for cognition.[15]

For a variety of reasons, including his carefully conducted experiments and monotheistic religious beliefs (which better comported with Christian rulers than the polytheistic beliefs of other Roman thinkers), Galen became the leading voice in the Western world's understanding of the brain and mind for well over one thousand years. There were, of course, scientific developments along the way. Leonardo da Vinci, for one, was able to use a wax-injection method to map animal brains and to see what ventricles actually looked like. But anatomical investigations on animals could lead only so far. For true advancement, the world needed to dissect human beings, which church authorities roundly denounced and largely blocked for centuries.[16]

Enter René Descartes, the great Enlightenment philosopher who famously concluded, "I think, therefore I am." Descartes was able to attend some human autopsies, despite church restrictions, and to conduct his own animal-brain dissections. In the center of the brain he noticed a small, pinecone-shaped organ known as the pineal gland. The gland is centrally located between the two hemispheres of the brain and close to the ventricles. This, Descartes figured, had to play a key role in directing human action. And the more he investigated, the more convinced he became that the pineal gland was the seat of the soul and the mind, responsible for directing spirits through the body.[17]

It was an important advancement based on empirical research, but it was nonetheless a reaffirmation of Galen's spirit theory. Thus, when Gall developed his theory, which he began calling *organology*, he proposed a series of ideas that upended thousands of years of scientific thinking, or what was known at the time as *natural philosophy*. His key insight was that the brain was the seat of the mind and that different parts of the brain performed different functions. Not a unified mass carrying out one task, then another, then another, the brain was actually composed of different organs that carried out different jobs and activated different mental functions, which could all be happening at the same time. As Gall summarized,

> The faculties and propensities of man have their seat in the brain. The faculties are not only distinct and independent of the propensities, but also the faculties among themselves, and the propensities among themselves, are essentially distinct and independent: they ought, consequently, to have their seat in parts of the brain distinct and independent of each other.[18]

Today we know this general idea as *cortical localization*, and though it has shifted somewhat from what Gall first proposed, it remains central to our understanding of the brain. To be sure, the other major part of Gall's theory—that the shape of the skull corresponds directly to the curvature of the brain underneath—is wrong, and modern neuroscience has shown that beyond a doubt. But science is always advancing in fits and starts; it's always improving, always incomplete. Gall did not have everything right in the late eighteenth century, but no scientist has it all right. Our current picture of the brain and mind may be better than Gall's, but it's certainly incomplete. Just as we look at organology with a bit of a smirk, our future ancestors will no doubt look at us with a similar smile.[19]

ONCE GALL WAS confident he had an insight to share with the world, he decided to publicize his work and to push his theory into the crucible of scientific debate that, he believed, would cement organology in the panoply of human knowledge. Yet he knew that his view of the brain and mind could rankle both scientific and religious authorities in Europe, so he began publicizing his work slowly and carefully, beginning with lectures and demonstrations out of his home in 1796. These presentations were small at first but grew steadily as word spread. Standing behind a table arrayed with skulls to illustrate shapes, masses, and bumps, he held up the remnants of the dead and unlocked the secrets of humanity for astounded audience members hungry for new knowledge.

After two years of homebound demonstrations, Gall decided it was time to reach a broader audience. But instead of penning some grand scientific paper, he chose to write a public letter to his friend Joseph Friedrich Freiherr von Retzer, secretary of the court and therefore part of Vienna's conservative political establishment. Though addressed to von Retzer, the letter was published in 1798 in *Der Neue Teutsche Merkur* (The New German Mercury), a literary periodical known for advancing Enlightenment ideas. This was the first written expression of organology, and it conveyed the basic principles of the science, Gall's method of investigation, and his findings thus far. And the letter worked—at least in the short term. Freiherr von Retzer was interested in the ideas, although he recognized they needed further refinement. Replying to the public letter, he told Gall to continue collecting evidence and formulating his theory.[20]

With Freiherr von Retzer's encouragement, Gall set about examining more heads, which became somewhat easier given the publicity his work was receiving. His friends in the medical field gifted him the brains of recently deceased people, which provided crucial insights about the ways brain structures corresponded to skull structures. It was also easy to obtain the skulls of criminals. After all, they were headed to hell once they died, so the least they could do was help the cause of organology. Obtaining the skulls of hospital patients was also straightforward for a doctor well connected in the medical world. Then there were people who committed suicide—more easy pickings for a well-to-do physician.

But Gall really wanted the heads of people who were not convicts, inmates, or unhinged. He wanted to test his theory on celebrities, statesmen, and leaders, who, unsurprisingly, were not inclined to hand over their skulls. So to examine the heads of Europe's movers and shakers, he turned to the next best thing—busts and casts that he could add permanently to his collection. These busts and casts, which would become stock-in-trade for subsequent phrenologists, were not exactly as enlightening as hands-on examinations of actual human skulls, but they were acceptable substitutes. Gall even managed to convince his friend Johann Wolfgang von Goethe, the famous poet, writer, and statesman, to sit for a cast of his head—a process Goethe detested.[21]

To add variety to his growing collection of European heads, Gall put out the word for people to send him exotic skulls from far-off lands. Eventually, he counted among his collection skulls from Africa, Asia, and the Americas—in fact, from every continent except Antarctica. But it wasn't just people heads he received; he also received the skulls of nonhuman animals—dogs, cats, birds, monkeys, apes, rabbits, deer, whatever he could get his hands on. It helped that he was a pet lover, and had been since a young age. Gall's house turned into a veritable menagerie, and whenever a pet died, another skull joined the collection.

How many skulls did Gall ultimately collect to support the new science? It's difficult to say with certainty. When he traveled on lecture tours, especially the long ones, he brought with him around 300 skulls, busts, and casts of people and animals. But this was just a portion of his larger collection. After he settled in Paris later in life, his collection was estimated at over 730 "pieces." This included the heads of young and old from infants to a 102-year-old man; the heads of the brilliant and the stupid—from idiots and imbeciles to purported geniuses; the heads of all types of criminals—highwaymen, murderers, thieves, and rapists; and the heads of the most famous people on the planet—Goethe, Friedrich Schiller, Franz Liszt, Voltaire, Francis Bacon, Descartes, and Napoleon, among many others.[22]

Gall even managed to collect his own skull. After he died in 1828 due to complications from a stroke, he insisted that his head not be buried with his body. Instead, a plaster cast of his head should be made, and his skull should be preserved alongside the other skulls in his collection. Today, a portion of his collection, including his skull, can be seen at the Rollettmuseum, just outside of Vienna.

ONE EVENING IN 1800, thronged by skulls, casts, and busts, Gall lectured at his home to Vienna's upper crust on the secrets of the human mind. As they listened with rapt attention, one member of the audience saw his future in the hollow, black sockets of the dead. He was a twenty-three-year-old medical student who earned extra money by tutoring the children of Viennese nobility. On that particular day, with his tutoring finally complete, he was able to settle into a comfortable chair in Gall's parlor and listen to a radical new theory. Gall's words set his mind ablaze.

After the lecture, the tall, dashing, sociable medical student strode up to Gall and introduced himself. His name was Johann Spurzheim, he said, and he wanted to learn everything he could about organology.

Gall looked at the young man and saw a promising brain underneath a high forehead, prominent brow, and thinning hair. He

Johann Spurzheim (1776–1832), Gall's handsome assistant and popularizer of phrenology across Europe

offered a standing invitation to Spurzheim to attend any lecture he wanted, free of charge.

Over the next four years, Spurzheim took Gall up on the offer countless times, and the two men developed more than a friendship. In 1804, Gall hired Spurzheim as his assistant, creating a kind of partnership that would, over the next decade, spread the new science across Europe. Eventually Spurzheim would overtake Gall in prominence and influence, even carrying organology across the Atlantic and capturing the minds of the young men at Amherst College.

Born in 1776 in a small German city near the Luxembourg border, Spurzheim showed signs of brilliance early on. His father, a farmer, saw his son's academic promise and began educating the young lad for a career in the ministry. By the time Spurzheim was fifteen, he had learned some Greek and Latin, but he had outgrown his hometown, so it was on to the University of Trier, where he learned Hebrew and excelled in philosophy and theology. In 1794,

French republican troops spread across southern Germany and occupied several towns, including Trier, and the university there was forced to shut down in 1798. After weighing his options, Spurzheim decided that his future lay in medicine, so he moved to Vienna, enrolled in medical school, and, a year later, met Gall.[23]

When Gall and Spurzheim officially teamed up in 1804, organology should have been thriving, but it had run into some powerful opposition. In the early years of the nineteenth century, the Holy Roman Empire, led by Francis II, was in trouble. Napoleon was conquering lands across Europe, even in North Africa and the Middle East. This rapid expansion exacerbated Francis II's already fraught relationship with France, which dated back to the revolution of 1793. Realizing his tenuous grasp on power, Francis II and his advisors decided to crack down on any potentially destabilizing forces, especially in Vienna, and Gall's revolutionary theory of the mind became an enemy of the empire.[24]

Even though Freiherr von Retzer, secretary of the court, had encouraged Gall to continue his research after reading the 1798 public letter, there were advisors closer to the emperor who believed Gall's ideas too dangerous for promulgation. One reason was because organology was supposedly materialistic, providing an empirical basis for humanity that eschewed the idea of God and undercut the need for religion, morality, and the church. Gall always insisted organology was not materialistic in the way his critics charged, but the criticism nonetheless plagued him until his death. Another reason was because organology affronted late eighteenth-century Viennese propriety. Specifically, the audiences for Gall's at-home lecture demonstrations included women and girls. These "promiscuous audiences," as they were known at the time, were taboo, if not downright dangerous to the moral order of society. There were even reports that Gall examined the heads of female audience members. Francis II and the Holy Roman Empire could not abide.[25]

So in 1801, at the behest of his advisors, Francis II issued an order that Gall cease his lectures immediately. Gall was taken aback by the order, especially given the number of other scientists and physicians at the time who lectured in their homes. Such lecturing was a standard practice, given that scientific equipment, visual aids, and artifact collections were often stored at home. Irked with the order, Gall petitioned the emperor to reconsider, denying that his ideas were materialistic, pointing out how many others lectured at their homes, and even offering to disallow women and girls from attending. None of it worked. In March 1802, Francis II ruled that medical lecturers could continue if they received the expressed consent of the dean of the medical faculty at Vienna, but he explicitly forbade Gall from delivering further talks.[26]

After a couple years in limbo, Gall decided that organology was too important to lay fallow. In 1805, he and Spurzheim packed up the skull collection and embarked on an expansive lecture tour that would last two and a half years and include stops in Germany, Denmark, the Netherlands, Switzerland, and, finally, France. The lectures made a decided impression on many attendees. As one listener recalled the scene,

> Gall entered the huge hall of the inn, surrounded by animal and human skulls. His lectures revealed his deep conviction, and he expressed himself with the ease of conversation. They impressed, and the comparison of the human skull with the animal skulls had something surprising. Thus, the skulls of infamous thieves were compared with those of (thieving) magpies.[27]

Spurzheim's role in the lecture hall was often to do the hands-on work, which meant holding and displaying specimens while Gall spoke. When addressing medical audiences, Spurzheim was tasked with dissecting a human brain to show the audience the cranial nerves, pathways to the cerebral cortex, and fibrous white matter. Over the years he became quite good with the dissection knife, slicing through gelatinous brain tissue with dexterous hands and a steely resolve. What he did "to human and animal bodies was not for the squeamish."[28]

In addition to lecturing, Gall and Spurzheim regularly visited prisons to study more heads. Happy to assist the proud organologists, prison officials marched their wards in a long line to an examination chair. In the dimly lit chambers of dank nineteenth-century dungeons, Gall and Spurzheim ran their fingers through unwashed hair, massaged greasy scalps, and recorded the masses and spans that landed these inmates behind bars.

They regularly visited mental asylums too. In Berlin, they visited a famous asylum, the Charité, and were accompanied by a cadre of magistrates and physicians. One by one they examined patients without knowing anything of the patients' background and madness. Gall and Spurzheim apparently did not make a single error in explaining the reason the person was in the asylum. In Amsterdam, they examined a madman who believed he was forced to sin and could do nothing about it. They found his organ of Devotion to be greatly developed. Also in Amsterdam they examined a woman who spoke incessantly of her pregnancy, although she was not pregnant. Her head was small, save for Philoprogenitiveness—the organ that controls love of offspring.[29]

At times, prison and asylum officials threw Gall and Spurzheim a curveball in order to test their ideas. At a prison in Germany, for instance, they examined the heads of around 250 inmates, most of whom were thieves. Head after head they felt, discovering basically the same shape around the areas related to cunning and larceny. But officials slipped in line a female inmate who was not a thief, just to see what Gall and Spurzheim would say. Feeling her head, they were taken aback. She was not a thief, no; this was more like the head of a murderer, they insisted. Prison officials were astounded; the woman had been locked away for infanticide.[30]

Gall and Spurzheim passed other tests with the same flying colors. The king of Prussia had stayed abreast of Gall's travels and lectures, and despite the stories he heard of Gall's abilities, he remained skeptical. So the king hatched a plan. He hosted a banquet for Gall and Spurzheim where a number of military officers were also guests. When the king asked them to examine the heads of the military officers, they happily obliged, but soon recoiled at what they found. These didn't seem like the heads of top military officials; the organs of Hostility and Destructiveness were too large, which Gall and Spurzheim pointed out. Smiling and nodding, the king had a confession to make: They weren't military officers at all, but convicts disguised in uniforms. As a token of esteem for passing the test, the king gave Gall a golden ring set with several precious stones.[31]

Throughout their travels, Gall and Spurzheim often managed to gain the interest and support of nonscientists. But scientists themselves usually had a different reaction. They enjoyed listening to the men—the organologists were great speakers and demonstrators—but they weren't particularly moved by the theory. In the Netherlands and Switzerland, Gall and Spurzheim were basically rebuffed by the scientific community and had to cut their visits short.[32]

In 1807, after two and a half years on the road, the now-famous organologists decided to settle in Paris. In truth, the political climate in Paris was not much better than in Vienna, given that Napoleon was not a fan of Gall's ideas; he thought that Gall was trying in vain to tame human nature. But Napoleon's wife was charmed by Gall and wanted to learn more about organology. So all things considered, Paris was the right place for them to enter a new phase of their work: publishing humanity's definitive guide to the brain and the mind.[33]

GALL HAD BEEN working on a big organology book for years by the time he arrived in Paris—from even before Spurzheim became his assistant. But with Spurzheim on board, the men were able to collect more data, record more examples, and feel more heads. Although the traveling they had done after leaving Vienna had slowed their writing considerably, it was in Paris that they could finally make the book happen. In 1810, the first of four volumes was ready for the printer.

The official title of the book was *Anatomie et physiologie du système nerveux en général, et du cerveau en particulier* (Anatomy and physiology of the nervous system in general and of the brain in particular). Over the next decade, three more volumes would appear, the final one being published in 1819. The volumes were shockingly expensive at the time, costing around $5,000 in today's dollars, but they were the complete statement of Gall's theory. The book detailed the functions of the brain, the work of the mind, the organs that shaped human behavior, and the correspondence between the skull and what was beneath. Finally, the years of toil, skull collecting, and battles with authorities had paid off.[34]

But behind the scenes, Gall and Spurzheim's relationship was fraying. The first two volumes of *Anatomie et physiologie* list both of them as authors. But the final two volumes list only Gall. In the process of writing the book, it became painfully clear that the men had fundamentally different views on the value of organology. Spurzheim was a deeply religious man—a liberal Protestant who was convinced that organology was the scientific balm for individual and societal ills. Gall, however, was a lapsed Catholic who wanted to keep organology strictly in the realm of science. His ideas—and they were his ideas, he insisted—needed further testing, rigorous experimentation, and collaborative research if organology were to join the ranks of other natural sciences.[35]

Beyond the religious and scientific differences, Spurzheim was an ambitious man desperate to emerge from out of Gall's shadow. For his part, Gall was not interested in sharing much of the credit for organology, despite naming Spurzheim his coauthor. Then there was the family issue. Spurzheim had recently married and needed to start making his own money, apart from the terms that Gall set for him. All these dynamics came to a head in 1813 with the appearance of the second volume when Spurzheim decided to move on from Gall. He packed his belongings and returned to Vienna to finish his medical degree (something he hadn't been able to do before leaving with Gall on the lecture tour). When the degree was complete, he and his wife moved to England.[36]

Because organology was still relatively young and closely tied to Gall, Spurzheim decided to rebrand his version of the science as *phrenology*, which means "discourse on the mind," a term he did not coin but did popularize. Notably enough, the first use of *phrenology* appears to have been by American founder Benjamin Rush, who coined it in a series of lectures in 1805 at the University of Pennsylvania. Rush was not describing what Gall was researching, but the idea of "discourse on the mind" or "science of the mind" was apropos of the kind of mental investigations then sweeping the scientific world. It was perfect for the kind of fresh start Spurzheim sought in England.[37]

Once settled in his new home, Spurzheim promoted phrenology anywhere and everywhere. A stellar public speaker and more affable and handsome than Gall, he offered different lectures at different times of day and in different locations to allow as many people as possible to learn of phrenology. He aimed his talks at popular audiences as well as at medical professionals, and because of his knack for marketing, he was able to draw sizable crowds of both. He also handed out free lecture tickets to reporters, who would, in turn, cover his talks in local newspapers. Even though there was a relatively strong anti-German sentiment in England at the time, Spurzheim had a commanding presence and exciting manner of presentation, and he quickly became a mover and shaker in the London social scene.[38]

Hearing of what Spurzheim was doing in England, Gall fumed with resentment. He hated both the word *phrenology* and Spurzheim's new "discoveries." In England, Spurzheim had expanded the number of organs of the brain from twenty-seven, which was what Gall had identified, to thirty-three, then to thirty-five, and ultimately to thirty-seven. These weren't huge alterations to Gall's theory; the new organs were typically just subdivisions of organs Gall had already named. But Gall took the expansion as an affront. In addition to increasing the number of organs, Spurzheim structured the organs in a more systematic way. Whereas Gall had simply listed the organs, Spurzheim grouped them into major categories:

Spurzheim lecturing with a phrenology bust, pointing to the organ of Eventuality

Feelings, which was subdivided into Propensities and Sentiments; and Intellect, which was subdivided into Knowing and Reflecting.

In 1815, Spurzheim published a book in England titled *The Physiognomical System of Drs. Gall and Spurzheim*, giving Gall top billing for the ideas. Yet Gall believed the book attempted to subsume his research into a new system. In lectures and subsequent publications, he charged Spurzheim with copying his original research, then corrupting it in a way that critics found easy to ridicule. With biting sarcasm, he remarked that Spurzheim was "very ingenious to have made a book by cutting with scissors"—effectively a copy-and-paste job from Gall's own books. In 1823, Gall went on a lecture tour of his own in the United Kingdom, hoping to correct the misconceptions Spurzheim had introduced. Unfortunately, audiences for the lectures were much smaller than Gall had hoped, and he soon returned to Paris, having lost the battle on Spurzheim's home turf.[39]

PART OF WHAT made Spurzheim so successful in spreading the word of phrenology were the friends and allies he made. Followers sprang up everywhere he went, but they were particularly ardent in Edinburgh, which would effectively become the hub of phrenology in Europe for decades. The de facto leader of the Edinburgh phrenologists was George Combe, a lawyer who was at first skeptical of the science but quickly became its most prominent advocate. Combe met Spurzheim in 1815 at a party and watched him dissect a brain in front of the half-soused guests. He also attended some of Spurzheim's lectures and resolved to keep the science going once Spurzheim left Edinburgh.[40]

In 1820, George Combe and his brother Andrew founded the Edinburgh Phrenological Society—the first of countless phrenological societies across the Western world. The brothers also advocated for the science ceaselessly, including in several influential writings. In 1828, George published *The Constitution of Man*, an instant classic that became the top-selling science text of the century, ahead even of Charles Darwin's *On the Origin of the Species*.[41]

With the Combes at the helm of phrenology in Europe, Spurzheim set his sights on the New World and the democratic revolution happening there. Phrenology's central message of human improvement—a message that Gall never fully trusted—seemed a perfect fit for a young, ambling, energetic nation trying to undo the influence of aristocracy and to put power in the hands of the people. That was never easy to do—an always incomplete effort, to be sure—but Spurzheim's assessment was apt. The new nation and the new science were made for each other. If only he had lived long enough to see them come together.

CHAPTER TWO

COMING TO AMERICA

THE MIGHTY *RHONE* SAILED FROM LE HAVRE, FRANCE, ON June 20, 1832. A packet ship constructed in 1831 to carry people, mail, and cargo across the Atlantic, the *Rhone* was designed to make the long journey as bearable as possible. But that was a tall order; six weeks at sea was unpleasant at best for the 150 people on board, most of whom crammed into small, working-class quarters. A majority of the passengers were French, but there were also people from the United States, England, Switzerland, and Germany. They listed their occupations on the ship's manifest as farmers, merchants, and manufacturers. At least one was a shoemaker. Then there were a dozen or so upper-class passengers, whose rooms were far superior to those of the working class. Their occupations were different—planters, lieutenants, and "gentlemen," which basically meant independently wealthy. Among this group was a man who listed his occupation as "Doctor." Johann Spurzheim may have found it easier than writing "Phrenologist."[1]

The doctor's training came in handy during the voyage, as he dutifully attended to the passengers who took ill. Years later, Spurzheim's friends recalled "his kind and successful medical assistance" to everyone who needed care. "Many of these poor men would perhaps have perished without your aid," a Parisian friend wrote to the doctor after the conclusion of the voyage. It may well have been a gratifying note to receive. For years the doctor had been maligned as a charlatan, but he knew, despite the opposition, that the science he practiced was a science of help, hope, and healing. That's exactly what he provided to people aboard the *Rhone*, and they were grateful.[2]

Spurzheim was happy to treat the sick aboard the ship, but his real attention was on his destination—the United States. The adolescent nation offered him endless opportunities to advance his science and to further his influence. In America he could examine the heads of American Indians and hopefully acquire some Indian skulls for his collection. He could study the heads of America's enslaved population, and he hoped for some of their skulls too. He could also spread the gospel of phrenology further and farther than it had spread thus far. By 1832, phrenology had its champions in the United States, but none of them were as charming, handsome, accomplished, and persuasive as Spurzheim.

Plus, his life in Europe had changed significantly over the past few years. Gall had passed away in 1828, and Spurzheim's wife had succumbed to illness in 1829. The time was right to go to America. He planned to stay in the country for two years, but he might extend that if everything went well. He could see himself thriving there for a long time.[3]

That wouldn't happen. When the *Rhone* finally docked in New York on August 4, Spurzheim had only three months to live.

SPURZHEIM STEPPED OFF the boat to a nation in the throes of something big, although no one at the time knew exactly what it was. In 1828, Andrew Jackson had defeated John Quincy Adams in the presidential election, a repeat contest of the 1824 election, which saw Jackson effectively robbed of a victory through backroom dealings in the House of Representatives. But 1828 set the situation right, with Jackson earning 178 electoral votes to Adams's 83 electoral votes. Especially telling were which states went for each candidate. Jackson carried fifteen states to Adams's nine, but all of Adams's states were small ones in the East—the old, well-established parts of the country. Jackson's states were the big, broad, sprawling states of the "West"—basically everything outside of New

England. The nation's center of power was moving away from where it had been since the founding. Amid Spurzheim's arrival in August 1832, another presidential election was right around the corner, and Jackson would extend his electoral victories, carrying 219 electoral votes to Henry Clay's 49 votes.[4]

Jackson's election signaled far more than a personal victory for himself, a war general often called "Old Hickory," a name derived from his tough, sturdy approach to life and battle. It signaled a nation ready for a democratic shift. The founders had rejected Europe's aristocracy of blood, where people were supposedly equipped to rule because of their lineage. But they had nonetheless embraced a different kind of aristocracy—what they called a *natural aristocracy.* Some people, they believed, were endowed by nature, not by family, to lead a nation. They were smarter than the others, with gifts of insight, virtue, wisdom, and leadership. In 1813, Thomas Jefferson made the idea explicit in a letter to his aging rival and pen pal John Adams:

> There is also an artificial aristocracy founded on wealth and birth, without either virtue or talents; for with these it would belong to the first class. The natural aristocracy I consider as the most precious gift of nature, for the instruction, the trusts, and government of society . . . May we not even say that that form of government is the best which provides the most effectually for a pure selection of these natural aristoi into the offices of government?

In turning against the artificial aristocracy of Europe, the founders sought to secure the natural aristocracy of republicanism. Of course, it was still an aristocracy.[5]

Then Jackson came along and called aristocracy itself into question, at least to his legions of supporters. No longer did the nation need aristocratic presidents from Massachusetts and Virginia, the home states of every president prior to Jackson. Jackson was born in South Carolina and lived in Tennessee, parts of the country that were wild, undeveloped, and bustling with frontier energy. Jackson's ascension to the White House promised to make the government more responsive to and representative of the frontiers. That's why, at his inaugural ball in 1828, over a thousand rough-and-tumble "ordinary" Americans showed up to the White House to celebrate, turning the occasion into something of a drunken riot (at least by some accounts) that damaged Jackson's new home and foretold a rollicking era of politics.[6]

Democracy was on the rise, which was why Alexis de Tocqueville crossed the Atlantic in 1831 to record observations for his book *Democracy in America*. Across the country, local officials who were once appointed by those in power were now elected by a majority vote (at least a majority of the people who could vote). Almost every state, save for South Carolina, now chose electors based on popular vote instead of legislative decision. At the same time, changes in printing technology, specifically the rise of the steam press, led to a proliferation of newspapers and periodicals, representing more voices, more points of view, more partisanship, and more confusion. The voice of the people was a cacophony—a rich, rough, uneven, democratic ballad to the world.[7]

The turbulence of the age led many Americans to see an opportunity at hand—a chance to correct the problems of the past and to put the nation on the right track. On January 1, 1831, William Lloyd Garrison, a bald, bespectacled Bostonian, published the first issue of *The Liberator*, which thrust America's original sin of slavery into the center of public life. Garrison pledged that his pen would serve the cause of immediate abolition "till every chain be broken, and every bondman set free! Let southern oppressors tremble—let their northern apologists tremble—let all the enemies of the persecuted blacks tremble." Garrison, who would become a friend of the Fowlers and an ardent supporter of phrenology, believed that the sin of slavery needed absolution if God was ever going to bless this nation and its democratic potential.[8]

There were other sins that needed absolution as well—specifically, personal sins of people who had veered from the path of righteousness. At the end of the 1820s and beginning of the 1830s, a religious awakening swept across the country—the Second Great Awakening, as it would come to be known. At the center of this revival culture was the charismatic, Holy Spirit–inspired preacher Charles Finney—another friend of the Fowlers—who held tent revivals across upstate New York. The enflamed passion of these revivals lit up the land, leading that part of the country to become known as the "burned-over district."[9]

As Garrison worked to break chains and Finney to save souls, Lyman Beecher worked to exorcise the demons of alcohol from American homes. Beecher, whose son Henry would team up with Orson Fowler at Amherst College to spread the gospel of phrenology, stood alongside Finney as the nation's leading religious figures. Looking around New England in the 1820s, Beecher was shocked by America's debilitating drunkenness and its effects on morality, family life, and democracy. How could the people rule themselves if they were all falling down drunk? In 1826, Beecher published a collection of six sermons on the evils of alcohol, naming it the most dastardly problem facing the nation. "Intemperance is the sin of our land," he thundered to his countrymen, "and, with our boundless prosperity, is coming in upon us like a flood; and if anything shall defeat the hopes of the world, which hang upon our experiment of civil liberty, it is that river of fire, which is rolling through the land, destroying the vital air, and extending around an atmosphere of death." The book, which was published by the American Tract Society, an evangelical publisher founded in 1825 to marry the Gospels with social reform, was a bestseller and helped kick off a temperance movement that would culminate in prohibition one hundred years later.[10]

These and similar movements came to define what historians call the "era of reform," which ran from around 1830 to 1860—the same decades that saw the rise of phrenology. Beyond abolitionism, religious revivals, and temperance, there were efforts to reform prisons and mental asylums; to give women an equal status in civil society; to provide elementary education for free to all American children; to communicate with the dead and receive the guidance of the spirit realm; and to spread new gospels, including one inscribed on golden plates and dug up in the burned-over district by a young man named Joseph Smith.

Equally as impressive as these social reforms were the scientific and technological advancements that seemed, at least to many onlookers, to overcome the limits of space and time. In 1804, a British engineer named Richard Trevithick designed the first steam-powered locomotive, although he was never able to produce a viable prototype. The idea, however, prompted engineers in other countries, including the United States, to devise their own engines. In 1814, the first successful steam locomotive made its debut thanks to the work of another British engineer, George Stephenson. Then, in 1830, Peter Cooper designed the first American-built steam-powered locomotive, known as Tom Thumb, which could travel upward of fourteen miles per hour—a breakneck speed when it came to carrying groups of people and crates of goods.[11]

In the mid-1810s, a French inventor named Nicéphore Niépce patented the basic idea of photography, although it took him around a decade to make it work. Niépce finally produced a lasting photograph in 1827, seeming to capture and preserve reality in a way heretofore impossible. Around the same time, inventors realized their visions for electromagnets, the microphone, the sewing machine, the typographer (a precursor to the typewriter), and a mechanical calculator that portended modern computers. One of the most important inventions of the era was Samuel Morse's telegraph, which started as an idea in 1830 and became a reality in 1837, when Morse demonstrated a working prototype in New York and Washington, D.C. A few years later, Morse sent a message from Washington, D.C., to Baltimore—actually, a question that captured the uneasy wonders of the age: "What hath God wrought?"[12]

In one way or another, these inventions signaled humanity's newfound power over reality, at least as it had been known for millennia. The telegraph collapsed space by making communication instantaneous, ushering in a new age of information. The steam locomotive moved dozens of people, animals, and goods across the country, making trips that used to take months in a matter of days. The other inventions, especially the ones involving electricity, seemed to erase the division between science and magic, sparking an era of wonder and amazement. The Industrial Revolution was born, allowing humanity to harness the power of nature for social progress. Into this dynamic environment of reform and improvement came a science that promised to unlock human potential and to make the world a better place, perhaps even a perfect place.[13]

THE DECADES-LONG SHIFT from artificial aristocracy to natural aristocracy to democracy was, in many ways, also the story of the Fowler family. They were "of the oldest Puritan stocks," tracing their lineage to William Fowler, who arrived in the New World from England in 1637.[14] Yet it was William's grandson, Jonathan Fowler, whom the phrenological Fowler siblings pointed to as their most significant forefather.

Known as the "Giant of America," Jonathan was almost seven feet tall and three hundred pounds of frontier muscle, a Paul Bunyan–type figure, but without a blue ox, whose life was handed down through a series of tall tales. Some of these tales involved the giant throwing casks of beer farther than anyone thought possible. Others involved him wrestling American Indians, disciplining troublesome slaves, and besting drunken Irishmen with one hand. There's even a story of Jonathan lifting the corner of a house right up off the ground. But the most repeated tales featured a bear and a shark.

According to colonial lore, Jonathan and a group of associates were in Canada when they came upon a massive grizzly. Startled from its hiding place, the bear rushed the group, snarling for blood and meat. The men fled like a bunch of cowards (or sensible human beings, as the case may be), except for old Jonathan, who stood his ground for battle. As the bear raised up on its hind legs, Jonathan seized it by the throat with one hand and, with the other, grabbed a knotty limb of pine. As his cowardly friends watched in amazement, he beat the grizzly to death with the limb. News of this feat soon made it back to England, where the king commissioned a painting of Jonathan, the Giant of America, draped in a bear skin. The painting hung in the palace for decades.[15]

On another occasion, in Guilford, Connecticut, Jonathan came upon a shark writhing around in a tide pool. It was a massive beast—five hundred pounds at least and pure, ocean-toned muscle. Seeing it struggle, Jonathan trudged out to the tide pool, slung the fish over his shoulder, and walked it back to shore through the sticky mud and swirling water of the coastline. It would provide him meals for days. "Quite a load," Orson Fowler later wrote of his ancestor, "for so slippery a commodity and bad a road."[16]

Jonathan's strength was a natural endowment, and he was a leader in the colonies because of it. Horace Fowler, one of Jonathan's descendants, was also a leader—or better, a pioneer of post-Revolution America who made his home in rural, unsettled Cohocton, New York. It was sparsely populated when Horace arrived there in 1808 (or 1806, by some accounts). Wildlife was plentiful in the area during the eighteenth century, providing a fertile hunting ground to several American Indian tribes. But white settlers quickly messed that up, clearing swaths of forest, redirecting streams, and pushing out wildlife. Nonetheless, Horace saw it as a great area to settle for his future family. Shortly after arriving, he met and married Martha Howe, and they set about making the land their own.

With Jonathan's strength running through his veins, Horace built a log cabin in the middle of a twenty-four-mile wooded area—"a wild and mountainous section of the country." In October 1809, they had a son, Orson. According to Fowler family lore, Orson was the first white child born in the area, although that was almost certainly incorrect. Local records indicate that Bethuel Hooker, another white child, was born there in 1800. Nevertheless, it was an important idea, allowing the Fowlers to see themselves as pioneers of the great American experiment. In June 1811, the family continued to grow with the birth of Lorenzo. Then, in August 1814,

Horace Fowler (1782–1859), the stern, devout patriarch of the Fowler family

along came Charlotte. Sometime after that, although the exact year is not clear, they had another boy, Theon.[17]

By most accounts, Horace was a prickly man. A prominent deacon in the local church, he took his role as religious leader quite seriously, especially when it came to discipline. When Orson did something wrong, Horace took him aside and asked him to choose his form of punishment. If Orson chose getting hit with a switch, Horace administered it, but first said a prayer aloud with the lad. That prayer, Orson later recalled, "had an infinitely greater effect than the whipping." It was an early lesson in the power of Veneration, a phrenological organ that was much better than physical pain at curbing vice.[18]

Lorenzo was apparently more of a problem child than Orson. A headstrong lad at only three years old, Lorenzo regularly cried, complained, and refused to follow directions. For a while, Horace chose not to punish the boy out of fear of setting him off on an epic tantrum. But that didn't last long. One day, Lorenzo was ordered to stop playing outside in the rain. He refused with a devious grin. Incensed, Horace grabbed the child, stripped him of his clothes, and dunked him in the family's rainwater cisterns. He let the little boy up momentarily to breathe, then pushed him under again. "You have been a very naughty boy," Horace screamed. "You don't pretend to mind me, and I intend to keep dunking you till you always do just as I tell you." Again and again he let the boy up for air, then pushed him back under. Gasping for life, Lorenzo finally promised to behave. After that, he was "the most faithful and obedient child in the family."[19]

Their mother, Martha, was as devout as Horace but not as mean. A saintly woman with the gift of prose, she wrote "manuscript by the bushel, and in a style at once graceful, flowing, perspicuous, and elegant." Unfortunately, Martha's literary abilities were paired with a sickly constitution. Her "high-wrought nervous temperament" combined in middle age with a diagnosis of consumption (tuberculosis), which took her life in 1819 at the age of thirty-six. Young Orson was only nine years old at the time, and he later recalled how losing his mother led him to crave "female sympathy without knowing it," an early lesson in hereditary bonds and character formation. Fortunately, there was a local Sunday school teacher, Mrs. Andrews, who poured her affection on Orson and "made me love her as if my mother." Later in life, he was able to examine Mrs. Andrews's head and found why she had been so caring: "Love large, I described her as a real missionary for good among young men, by virtually adopting them in feeling, and molding them."[20]

As was typical in the nineteenth century, Horace quickly remarried after the death of his first wife so that his children and household had a motherly influence. His second wife was Mary Taylor, and surviving accounts indicate that Mary was a suitable figure for the kids—kind, pious, and a natural educator. Horace and Mary had four children of their own—Samuel, born in 1821; Martha, born in 1823; Almira, born in 1826; and Edward, born in 1831. Each member of this second group of Fowler children would support the cause of phrenology in one way or another.[21]

As Orson progressed into his teenage years, he came to realize that he inherited from his mother something besides her literary gifts—namely, her sickly constitution. His lungs were his biggest problem. When he was fifteen years old, he developed difficulty breathing and had to remain indoors for three months, which ended up propelling his bookish, literary passions. Respiratory problems would stay with him most of his life, wreaking particular havoc on his college experience and forcing him to restart his junior year because of another season of difficult breathing.[22]

With a younger brood of Fowlers to contend with, Horace knew it was time for his eldest sons to leave the family homestead and to embark on their own course in life. Devout as the day was long, he wanted nothing more than for Orson and Lorenzo to become ministers, and a college education was the best way to do that. Orson had his sights set on Amherst College, which had welcomed its first students in 1821 and would be an ideal place to learn to preach the Gospel. The big trouble was simply getting there. With only four dollars in his pocket, Orson summoned the strength of the Giant of America and walked over four hundred miles to enroll at Amherst.

Once enrolled, he paid his bills by sawing, splitting, and carrying wood for his classmates. He often had to climb three or four flights of stairs with heavy cords of wood in his arms. Apparently his lungs were up to the task. The other students paid him and laughed at him—the rube from the backwoods of New York hauling their stuff. But Orson knew the work strengthened him (at least until his respiratory problems caught up with him). Physical labor combined with intellectual pursuits was key to success in life, and it yielded a "rich harvest" for Orson personally and for phrenology more generally.[23]

ALTHOUGH SPURZHEIM'S VISIT to the United States put phrenology on the map, word of the new science had made it to America in dribs and drabs since the early 1820s. The earliest references came via newspaper reports from overseas including a lengthy reprint of a European article describing a recent meeting of the Edinburgh Phrenological Society. Then, on July 22, 1822, the *Cincinnati Gazette* announced an upcoming series of lectures delivered by "Dr. Cranium" on the "highly interesting and novel science of phrenology." The lectures, which were to be delivered by a man whose name was too on the nose to be real, would take place at the Lunatic Asylum of Ohio and defend the science from the "shafts of ridicule and the arrows of malevolence" that sought to "retard" it. In addition to vanquishing phrenology's foes, Dr. Cranium made sure to include one important detail about his connection to the science—namely, that he had recently been in Europe and had "the supreme felicity of an introduction to the immortal Spurzheim, who was then lecturing upon Phrenology before the Royal Institute of Paris." Apparently Spurzheim had felt Dr. Cranium's cranium and had deemed it "particularly well formed for a successful lecturer on the science."[24]

Dr. Cranium was one of several Americans who heard Spurzheim lecture in Paris and returned to the United States excited to propagate the new science. Another was the Harvard anatomist John Collins Warren, who attended one of Spurzheim's lectures in 1821, then returned to the United States and began speaking on phrenology to other members of the Harvard community. Still another was John Bell, a prominent Philadelphia physician who spoke about it around the Northeast.[25]

But the most important early advocate of phrenology in America was Charles Caldwell, whose work would become infamous for its proslavery import. Caldwell is a challenging figure in the history of phrenology. On the one hand, he was an early popularizer of the science, a vigorous lecturer and writer who understood how to adapt Gall's system to different people and causes. On the other hand, he is a poster child for the charges of scientific racism that followed phrenology into the twenty-first century. A proud slaveowner, he maintained that phrenology proved the supposed superiority of the white race. It's even said that Caldwell was the inspiration for the Leonardo DiCaprio character Calvin Candie, an unapologetic slave-owning phrenologist, in the Quentin Tarantino film *Django Unchained*.[26]

Charles Caldwell (1772–1853), prominent phrenologist and stalwart white supremacist

Born in North Carolina and determined to become a respected professor of medicine, Caldwell earned a medical degree from the University of Pennsylvania at the end of the eighteenth century, having studied under Benjamin Rush. Lean, balding, and angular, with a prominent Romanesque nose, a round, robust forehead, and a sturdy brow, Caldwell may well have heard the term *phrenology* from Rush, but he didn't "convert" to the new science until he traveled to Europe in the early 1820s and listened to Spurzheim lecture in Paris. A trained physician, he saw the incredible advancements in brain science that phrenology represented, and he praised the way it mediated between "pure spiritualism" and "pure materialism," both of which were unworkable extremes. In his mind, the promises of phrenology were clear from the outset. "If the science be true"—and of course he believed it was true—"its practical utilities are manifold and great," perfectly fitted to "amelioration of the condition of man."[27]

After learning of phrenology in Europe, Caldwell returned to the United States ready to start a scientific revolution. Along with John Bell, he founded the first phrenological organization in America, the Central Phrenological Society in Philadelphia, which aimed to lift the new science out of the realm of ridicule. "The study of phrenology," Caldwell and his compatriots announced in the *National Gazette* in March 1822, "is now assuming an importance and a philosophic character, which will require more from its opponents than the hackneyed exclamations . . . which ignorance can so readily utter, but which reason and cool reflection soon discover to be as unjust as illiberal." Despite the auspicious start, the organization didn't thrive as its supporters had hoped. Due to lack of interest from all but a handful of physicians and scientists, the Central Phrenological Society closed its doors in 1827.[28]

Caldwell was actually not in Philadelphia to see it close down. He had already taken a job in Lexington, Kentucky, at Transylvania University, where he was charged with creating a world-class medical school. Of course, a big part of the school's curriculum was phrenology, which Caldwell taught to students as soon as they arrived on campus. In 1824, his lectures to medical students were collected and published under the title *Elements of Phrenology*, which became the first U.S. book on the science. He also traveled on expansive lecture tours across the South to promote phrenology. To many Southern audiences (and Northern ones besides), phrenology provided pretty clear evidence of the superiority of the white race.[29]

Over the next few years, Caldwell developed a reputation as the nation's leading proponent of phrenology, although his success would be dwarfed by what the Fowlers would later accomplish. In Baltimore he defended the science against criticism that it was materialistic, fatalistic, and irreligious. In Washington, D.C., and Philadelphia, he explained it to statesmen, business leaders, and prominent thinkers. In Cincinnati, his lectures led to the creation of a local phrenological society, which has the distinction of being the quickest society to open and then shut down. The first meeting of the Cincinnati society attracted around twenty citizens, but the group shrank considerably for the second meeting. At the third meeting, when dues had to be paid, only the treasurer showed up. That was the end of it.[30]

Of all the parts of the country where phrenology tried to take root, the most fertile soil proved to be in Boston. The science had started to spread around New England after John Collins Warren had returned from listening to Spurzheim lecture in Paris, and it quickly intrigued scientific men in the area. Not all of them were fully persuaded by it, of course, but they believed it warranted further study. Nahum Capen was one such person who facilitated its growth through his firm Marsh, Capen, and Lyon, which was the leading publisher of phrenological books for a decade. In addition to phrenology, Capen wrote extensively on law, politics, and culture, imbuing the science with a democratic, American ethos that meshed perfectly with the Fowlers' subsequent advocacy.

As reports from overseas, particularly from the Edinburgh Phrenological Society, continued to appear in American periodicals and newspapers, the name Spurzheim acquired a halo that captivated Boston phrenologists. In the late 1820s they began trying to lure the world-famous head reader to the United States, but Spurzheim deferred on several occasions. Finally, in 1832, the siren song of the New World proved too alluring for Spurzheim to resist. In America he could probe the "genius and character of our nation," as Nahum Capen summarized. He could also examine some of America's distinct specimens—namely, American Indians and enslaved Africans. In accepting Capen's invitation, Spurzheim hoped he would have the chance to visit one of his favorite philosophers, William Ellery Channing, the transcendentalist, liberal theologian, and Unitarian minister. Channing's religious vision and theological innovations thrilled Spurzheim, and the possibility of meeting him in person was too exciting to resist. "Shall I not see Channing?" he asked his friends when discussing his plans. So Spurzheim boarded the mighty *Rhone* for the New World.[31]

DURING THE SUMMER of 1832, New York was pulsating with disease—specifically cholera. It was the first such outbreak in the New World, and the epicenter was close to the port where Spurzheim disembarked. "The largest and filthiest, the most crowded and vice-disfigured of American cities," New York treated public health concerns with only a passing thought. The streets of the densely packed urban jungle included thousands of swine that scavenged for food and left their mess behind. Piles of garbage amassed in gutters, waiting for municipal workers to collect them, which often didn't happen for long periods of time. Even worse was the city's water system, which was jokingly considered the best in the nation because it could clean dishes and flush out your insides. The cholera epidemic of 1832 left trails of diarrhea, vomit, and death from Battery Park to what would later become the Upper West Side, with the worst of it in Five Points in Lower Manhattan. All told, around 3,500 people in New York died from the disease, out of a total population of around 250,000.[32]

Spurzheim and the other passengers of the *Rhone* had to quarantine for a week before they could leave the city. Once able to depart, he headed first to New Haven, Connecticut, where Yale College was having its commencement week. The crisp, clean air of New England greatly improved his health, and he was able to mingle with many Yale faculty members. "The Professors were in love with him," recalled the noted chemist Benjamin Silliman. Spurzheim even watched the commencement exercises and attended a meeting of the Society of the Alumni, where he dissected the brain of a child who had died of hydrocephaly. His skill at slicing through delicate brain tissue greatly impressed even the most stoic medical men in the audience.

After New Haven it was on to Hartford, where Spurzheim visited his first American asylum. Stopping at both the Connecticut Asylum for the Education and Instruction of Deaf and Dumb Persons and the Connecticut Retreat for the Insane, he examined heads, described characters, and wowed those in attendance, including

Amariah Brigham, who would soon become a dedicated phrenologist and lead several mental-health reform efforts. With Spurzheim's influence and Brigham's passion, phrenology would usher in a new era of care for those who populated the nation's asylums.[33]

Finally on August 23, Spurzheim arrived in the city that made his trip possible—Boston. His boarding house for the duration of his stay was Mrs. Le Kain's on Pearl Street, which offered the finest dwellings for travelers then available in the city. Pearl Street was the perfect spot for his stay—a part of the city alive with ideas and possibilities. Nearby was the Boston Athenaeum, where he could listen to and deliver lectures on the newest, most exciting thought of the times. There was also the Perkins Institution for the Blind, which would host Spurzheim on many occasions in the coming months. Then, of course, there was Spurzheim himself; his presence shifted the attention of Boston's intelligentsia to Pearl Street. "The rich and the learned, the student and the scholar," recalled Nahum Capen, "soon paid him their respects, as due to a distinguished stranger, and a course of polite engagements was at once commenced. All who called upon him soon became his admirers, and he was made the leading topic of the day." *The Boston Medical and Surgical Journal* was also bullish on what his presence meant for the city: "The efforts of Dr. S. will form among us a new era in education, and open, to the minds of the most intelligent, new and correct views of their moral and intellectual powers."[34]

Shortly after his arrival, Spurzheim began delivering lectures to the people of Boston. His first venue was the Atheneum, but its auditorium proved too small to accommodate the eager crowd who showed up. So organizers hastily moved the talk to the Masonic Temple. He lectured on phrenology practically every day of the week, drawing sizable audiences who were hungry for his revolutionary science. In short order, not only was the word *phrenology* known around the city, but the basics of the science received extensive attention in local papers. On August 30, *The Boston Morning Post* ran a front-page, two-column article that recounted a brief history of phrenology's developments, explained its basic principles, and provided a summary of each phrenological organ. All told, the paper concluded, phrenology "has, and will continue, to benefit and improve every branch of practical knowledge, not even excepting philosophy, medicine, education, or legislation."[35]

During the day, prior to his evening lectures, Spurzheim often visited different institutions in and around Boston—asylums, hospitals, prisons, and schools—to feel the heads of whomever he could. He was particularly excited to visit schools, which signaled the kind of democratic promise that brought him to America in the first place. At Boston's Monitorial School, Spurzheim walked around a classroom, talked to kids, and examined them. One of the heads he felt was that of a five-year-old girl, about whom he remarked: "Fun, fun. Courage, too. Look out for pranks!" The teacher was impressed. The girl had been in his class for only a few days, but she had "already exhibited symptoms of insubordination," the teacher recalled. Spurzheim also visited the Hancock School and the Smith School for Colored Children, where he relished the chance to examine African American students. In general, he noted that the faculties of Individuality and Eventuality were strong among them, but their reflective faculties were less developed than those of white children. He also insisted that Black kids can often learn faster than white kids when young but that they will eventually lag behind white learners in their teenage years.[36]

When not lecturing or touring local institutions, Spurzheim hobnobbed with New England's luminaries. He socialized regularly with Harvard faculty, including president Josiah Quincy and famed scientist Nathaniel Bowditch. He also kept company with Boston mayor Charles Wells and national political celebrity Daniel Webster. He was particularly interested in conversing with prominent clergymen, including Edward Beecher, Hosea Ballou, and the eccentric "Sailor Preacher" Father Edward Taylor. Unfortunately, the famous Unitarian minister William Ellery Channing was too ill to visit with Spurzheim during his brief time in the United States. But Channing did later reflect on Spurzheim's influence: "He made a greater impression in our city than any foreigner . . . All who knew him here were struck with his unaffected philanthropy, and this was the great charm of his lectures."[37]

Spurzheim's life in Boston quickly settled into a rather demanding rhythm, with "hardly an hour in the day, after 9 o'clock a.m., during which he was not engaged either in receiving company or making visits." From the outside, his work was a smashing success, and newspaper coverage was effusive. But beneath the surface, Spurzheim was getting weaker and sicker with each passing day. At the end of October, audience members for his lectures saw him shivering violently as he tried to speak. On one occasion, after he had spoken on the natural language of the body, he quipped: "I feel quite ill, and I am afraid my own natural language has been too strong for the pleasure of my hearers." Friends tried to persuade him to postpone his ongoing course of lectures and to take time to recover, but he was pigheaded and overly confident in his ability to push through the pain. "The arrangement has been made," Spurzheim insisted. "The public will expect to hear me at the stated time, and when I have finished, it will be a relief to know that I can rest without disappointing others." The next evening, October 29, he showed up to lecture at the Masonic Temple and was more ghost than man. His "usually lighted up" appearance was "grave and feeble." Sweating profusely, drained of all color, and shaking from the chills, he nonetheless delivered his lecture—the last one of his life.[38]

The next day, Spurzheim could not get out of bed. Eminent local physicians visited him to see what could be done, but his fate seemed sealed. In a few days he lost the ability to speak and the urge to eat, and friends had to help him sit up in bed. Then, after ten days of drifting in limbo, Spurzheim passed away "without a groan or a struggle." The time of death was eleven in the evening on November 10, 1832, the official cause being typhoid fever related to the cholera outbreak that had greeted him upon his arrival in New York.[39]

In announcing his death, newspapers were in awe of what he had accomplished during his life and short time in the United States. "Dr. Spurzheim's head," wrote the *Boston Daily Atlas*, "is one of the finest that possibly could be selected to sustain the doctrine to which he devoted his whole life." *The New-England Galaxy* wrote, "This distinguished man fell a victim to his own zeal for the promotion of science." The *Boston Daily Advertiser* called on citizens to show Spurzheim the respect he deserved for a life dedicated to science and the improvement of humanity: "His death is sincerely lamented as that of an amiable and accomplished man. We presume that our citizens, in committing his body to the earth, will not do it without some demonstration of respect for the memory of a learned stranger, who has thus closed his days among us."[40]

ON NOVEMBER 17, the bells of the city of Boston rang continuously from two to three in the afternoon. Then Spurzheim's funeral began at Boston's famed Old South Church. At capacity long before the service commenced, the venue had to turn away many Bostonians who wanted to pay their respects. Ultimately, an estimated three thousand people were present to witness the prayers, poems, and orations that honored the great scientist. As the *Boston Daily Advertiser* summarized the occasion, "The final tribute of respect was paid to this distinguished stranger by a multitude of our citizens, whose respect and regard he had conciliated by his scientific reputation and the amiable qualities of his private character."[41]

When the funeral was over, a procession of several hundred people—"the most illustrious men of Massachusetts"—escorted Spurzheim's body from the Old South Church to the "receiving tomb" under Park Street Church, where it was embalmed. There it sat for a few days until Winslow Lewis and a few other physicians carried out Spurzheim's final wish, removing his skull from his body and preserving it for the future of phrenology. Carefully and reverently, Lewis sliced through Spurzheim's neck, detached his head, and scooped out his brain, which was placed, alongside the heart, in a fireproof case. Then they did with the head what was necessary to preserve the skull—boiling it to remove excess flesh and to brighten the bone beneath.[42]

With Spurzheim's skull, brain, and heart preserved, his headless

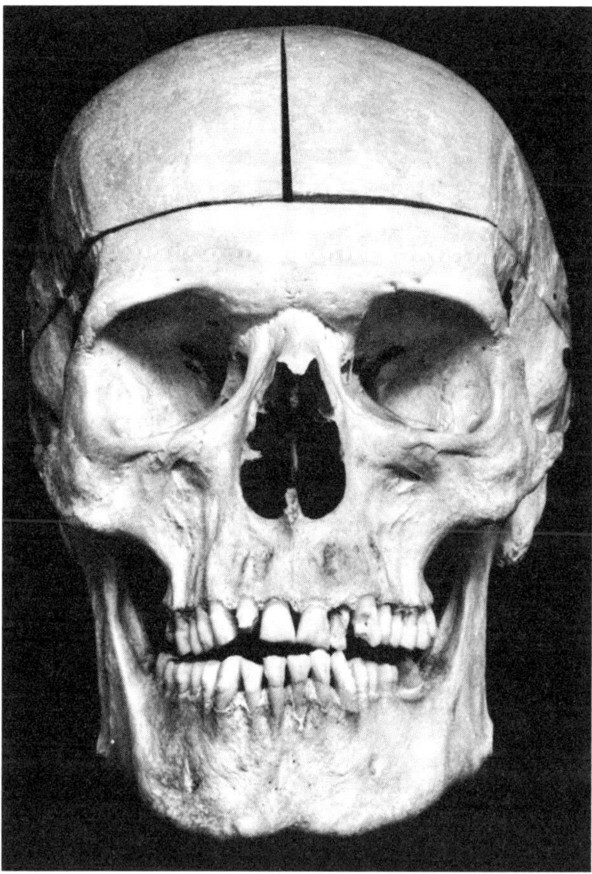

The skull of Johann Spurzheim, added to his collection shortly after his death

corpse was taken to Mount Auburn Cemetery, where it was laid to rest under a resplendent monument made of Italian marble, paid for entirely by the wealthy merchant William Sturgis. Befitting the reputation and influence of the man whom the monument recognized, only one word was carved into the marble: Spurzheim. That's the only thing visitors at Mount Auburn needed to know; his life spoke for itself.

To handle Spurzheim's belongings, including his specimen collection, his friends created the Boston Phrenological Society. The leading such society in the United States for many years, it catalogued and preserved Spurzheim's impressive collection of over four hundred skulls, casts, and busts. Perhaps the most important artifact of them all was Spurzheim's own skull, which was the last piece added to the collection. (Today, Spurzheim's skull is preserved in the Warren Anatomical Museum at Harvard University.)[43]

At first, the future of phrenology in Boston looked bright. The Phrenological Society counted among its ranks prominent physicians, clergymen, businessmen, and statesmen. But after years of trying to push the science forward, the society foundered. None of the local movers and shakers who originally signed up had the energy, charisma, or entrepreneurial vision needed to spread a new scientific gospel. That gospel needed its own John the Baptist, who, in 1832, happened to be one hundred miles away at Amherst College.

CHAPTER THREE

HEADS UP

THE BOOKS WERE BOUND WITH A HIGH-QUALITY, RED-patterned leather that looked almost like snakeskin, designed to protect the printed word at a time when books were scarce treasures. These particular books were not only scarce but hot off the press and shipped from the publisher in Boston directly to Beecher in Amherst. With the books, the American editions of foundational phrenological works, including those by Spurzheim, Beecher went from being a skeptic of the science to one of its most eloquent supporters.

When Orson Fowler borrowed the books from Beecher after the Amherst debate in 1832, young Lorenzo Fowler had only a vague idea of their significance. He had heard his older brother talking about phrenology, but these richly detailed tomes went beyond schoolboy talk and unlocked the secrets of the human mind. Equipped with a head for serious inquiry, Lorenzo dove into the works of Gall and Spurzheim as though they were sacred texts harboring a new gospel.

It was fortunate that Lorenzo had moved in with his older brother right around the time Orson and Beecher teamed up. The younger Fowler brother had left home at seventeen to attend Dansville Academy and to prepare for a career in the ministry. While attending school he lived with a local minister, Rev. Hubbard, who sponsored and tutored the young man until he was ready for more rigorous schooling. When that time came, Lorenzo relocated to Hadley, Massachusetts, to complete his secondary studies and to prepare for matriculation to Harvard. But fate and finances intervened before he could enroll, sending him in 1832 to live with Orson in Amherst. The college there would be a fine runner-up to Harvard, Lorenzo concluded, and both he and Orson could graduate as ministers from the same institution. But then phrenology grabbed hold of them.[1]

Lorenzo watched in awe as Orson and Beecher examined the heads of fellow students, charged money for their services, and delivered lectures in neighboring towns. All the while the younger Fowler pored over books, busts, and crania with a religious purpose befitting a minister in training. When Orson graduated in 1834, the brothers were at a crossroads. They still wanted to honor their father's wishes and enter the ministry, despite the money Orson was making with phrenology. There was a strong possibility that they could matriculate to Lane Seminary in Cincinnati, Ohio, where Beecher's father, Lyman, was the president. But venturing to Ohio and enrolling at Lane would take time and money.

Orson then heard about one of his classmates delivering a series of lectures in Brattleboro, Vermont. The lectures were titled "Battles of the American Revolution" and, according to Orson, simply rehashed what he and his classmates had learned at Amherst. The talks were apparently a disaster, but they gave Orson an idea. One night in late summer 1834, he lay awake "till broad daylight" considering the possibility that lecturing was his true calling and not just something he did with Beecher during college. In the morning, he ran the idea by Lorenzo, who thought it brilliant.[2]

To prepare, Orson and Lorenzo began to study phrenology even more intently. They purchased more books on the science and a model phrenology head from Marsh, Capen, and Lyon in Boston. Then they immersed themselves in study. Right around this time, Orson realized something was missing from the way European phrenologists practiced the science. Specifically, they had no system for accounting for relative degrees of development. In response, he devised a phrenological chart that included three possible degrees—small, medium, and large—for every organ. This, Orson later boasted, was "the first attempt at such degree description" anywhere in the world. With the new chart carefully designed, Orson bought paper, hired a printer, and produced one thousand copies. Then he hit the road, leaving Lorenzo at home to finish his education.[3]

Following the lead of his Amherst classmate, Orson began his career as a professional phrenologist in Brattleboro. His initial lecture went well, and afterward he examined scores of audience heads, charging men 12.5 cents for a marked chart of their character, and women and children 6.25 cents. He pulled in $40, equivalent to the princely sum of around $1,400 today. After Brattleboro, it was on to Saratoga, New York, about seventy-five miles away, where he did it all over again.[4]

Yet after only a few lectures, Orson realized that working with a partner was a lot easier than speaking solo. So he turned to the person he knew could do the job and invited Lorenzo to meet him on the road. Without a second thought, the younger Fowler packed his belongings and headed out, connecting with Orson in Waterford, New York, and forming their first official partnership. After Waterford they hit Troy, Lansingburgh, Catskill, Amsterdam, Schenectady, and Hudson. In each locale, they lectured, examined heads, and met a colorful cast of characters. There was the woman who could communicate directly with God and serve as his mouthpiece here on earth; her Marvelousness and Veneration were very large. There was the man who was just the opposite—"incapable of being affected by anything bordering upon the supernatural"; his Marvelousness was very small. There was the young artist who could draw perfect portraits of people after only a brief glance at them; his Constructiveness, Imitation, and Perception were all quite large. Then there was the young woman with large Ideality, Comparison, and Language; the brothers concluded that she must be a talented poet—which she was.[5]

Additional "hits"—times when the phrenologists accurately captured the character of their sitter and revealed somc important truth—came fast and furious. In one city, although he never specified which, Lorenzo was walking down a dusty thoroughfare and passed a blacksmith's shop. Hearing the piercing clang of metal on metal, he looked inside and caught the eye of a blacksmith's assistant named Ray. Ray instantly recognized Lorenzo as the traveling phrenologist who had been in town for several days, and, despite being covered in sweat and soot, called out: "Please to put your hand on my head, and tell me if I must work all my life here!"

Lorenzo agreed. What he found was fascinating. Ray had a natural talent for inventions, Lorenzo could see, and told the laborer to become an entrepreneur.

Ray chuckled and dismissed the idea. What did he know about inventing things? Amused and unconvinced, he returned to the hammer and anvil inside the shop.

A few years later, Lorenzo ran into Ray once again, but this time the former blacksmith's assistant was dressed like a proper gentleman. After exchanging pleasantries, Lorenzo asked him what had changed.

"Ah," Ray replied, "I have taken out thirty-three patents, and each one does just what I intend it shall. The last is to make India-rubber springs for railway carriages, and these are now introduced all over the world. I have now 500 men to work for me, and have to thank Phrenology for calling my attention to the idea that I could be something besides a day-laborer."[6]

Giving a young man hope, direction, and purpose in life—that was the promise of practical phrenology that kept the Fowlers moving from town to town. Of course, they made decent money along the way, but they were propelled by a sense that they were riding the crest of a wave that would soon flood the world with knowledge. That's certainly what got them excited about their stop in Hudson, New York, where the brothers got to see firsthand how phrenology could inform the treatment of insanity.

The Hudson Lunatic Asylum, built in 1830 about thirty miles south of Albany along the Hudson River, afforded its fifty patients a resplendent view of the Catskill Mountains. Although located in the center of town, it was surrounded by lush gardens and a spacious area for exercise. Its sixty windows—thirty on the front of the building, thirty on the back—allowed cool mountain breezes to sweep in off the river, providing much-needed fresh air to the dozens of people struggling to get better. One-half of the asylum was reserved for women, the other half for men, but for both populations every care was "taken that nothing should offend the eye or disturb the imagination." It was to be a site of "comfort and restoration," a counterpoint to the other asylums of the era that had become places of pain and degradation.[7]

Orson and Lorenzo hoped to see as much when they toured the facility in the fall of 1834, escorted by the superintendent, Samuel White. Actually, White first asked the brothers to examine his own head, which revealed his great Benevolence and concern for others. Satisfied with the exam but still curious about phrenology's claims, White set up a test. He would seat a patient in front of them, and they would have to feel the patient's head, describe his or her character, and ultimately indicate why the patient was in the asylum in the first place. The Fowlers were happy to play along.

One patient they examined was a young man whose Ideality, Constructiveness, Imitation, and perceptive faculties were all very large. This man must be an artist, the Fowlers surmised. That was right, White indicated. The young man was so obsessed with fine art that he clamored constantly to visit Italy. In fact, his mother had to restrain him from fleeing the United States, which produced the "partial insanity under which he then labored." But at least at the asylum the young man was able to paint and pursue the creative activities that drove him. He even showed Orson a "beautiful and accurate specimen of miniature painting" he had produced.[8]

The Fowlers then examined a man with very large Combativeness and Destructiveness. "Sullen and fierce, and subject to violent out-breakings of passion," the man must have been a threat to public safety. Another hit, White indicated. Then there was the elderly lady with a similar constitution, although she also had a large organ of Language. Orson knew that was an interesting combination and figured that she was likely a threat to public safety as well, but also had a poetic streak in her. Hit again, White acknowledged. Her violent outbursts were "a turbid torrent of abusive eloquence that might have passed for prize-speeches in the hall of Pandemonium."[9]

In addition to examining patients at the asylum, the Fowlers examined heads at people's homes—a kind of phrenological house call. It was more decorous for the brothers to visit private homes than for women and children, unaccompanied by men, to venture out in public. House calls were also convenient for men who wanted their whole family examined. That's what happened with a local butcher in Hudson. Orson went to the butcher's home to examine his brood and found some rather shocking heads. One of the butcher's kids had extremely large Destructiveness and relished the sight of cattle being slaughtered. This was, in fact, his favorite pastime, and he objected vehemently to having to sit through a phrenological exam while there was an ox being slaughtered just down the street. The boy calmed down only when his parents assured him another ox would soon be killed. Then there was the butcher's three-year-old son, who had grown up around slaughter and loved it with a kind of maniacal glee. With Destructiveness very large, the bloodthirsty toddler had recently caught a small pig in the street and attempted to slit its squealing little throat with a dull knife before a passerby intervened and put a stop to it.[10]

Orson recounted this visit to the butcher's home as an example of the kind of behavior that results from and perpetuates overly developed mental organs. These kids were immersed in the brutality of the slaughterhouse, their minds coming alive at the sight of blood and the cries of animals. But hope was not lost for the family. Though Orson did not comment on how he helped to rectify the situation, he no doubt gave the butcher and the butcher's wife some practical suggestions for setting the kids straight. He likely told them to limit the kids' exposure to blood and death, to cultivate their Cautiousness and Benevolence through particular mental exercises, and to get them to church more often to enhance their Veneration and Sublimity. At its core, practical phrenology was a program for improvement, and children were never too young to take the first step down the road of health, happiness, and harmony.

In city after city, town after town, the Fowlers moved in and out of fascinating scenes of American life. Mundane and spectacular, virtuous and vicious, bloody and beautiful, they were the scenes one would expect from a young nation propelled by democratic energy and momentous problems. The more towns they visited, people they examined, and glimpses of life they secured, the more convinced they became that phrenology was the key to making sense of America's promise and peril. Their future was not as ministers of Christ; they were called to preach a new gospel.

WHEN PEOPLE WENT to see the Fowlers for a reading or had the brothers visit their home for a private examination, they came away with a potentially life-changing piece of paper—a chart that indicated what kind of person they were. The chart listed all of the phrenological organs and their respective sizes, giving Americans a quantified, objective version of themselves.

Practical phrenology was built on a grand taxonomy of the brain and mind, which included dozens of "organs" grouped in a deliberate, structured system. At the highest level, there were two orders, which were then divided into genus and species, then into specific organs. The organs were grouped on the head according to their function. In general, the farther down and to the back of the head the faculties were located, the more base and animalistic their role. Conversely, the higher up and to the front of the head the faculties were located, the more ennobling and distinctly human they tended to be. In general, the Fowlers maintained that there were thirty-seven different organs, which was the number that Spurzheim settled on late in his life. The brothers provisionally added new organs to the list of thirty-seven, which they indicated with letters instead of numbers so as to distinguish them from the Spurzheim-approved organs. With everything put together, the Fowlers' system of phrenology looked like this:[11]

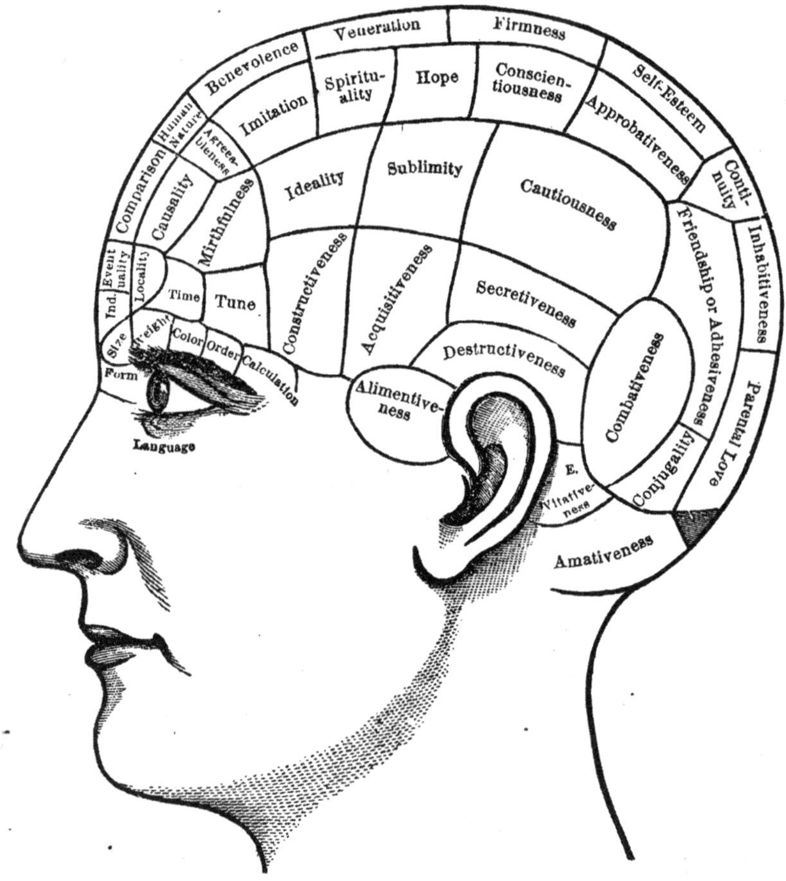

NUMBERING AND DEFINITION OF THE ORGANS.

1. Amativeness, Love between the sexes.
A. Conjugality, Matrimony—love of one. [etc.
2. Parental Love, Regard for offspring, pets,
3. Friendship, Adhesiveness—sociability.
4 Inhabitiveness, Love of home.
5. Continuity, One thing at a time.
E. Vitativeness, Love of life.
6. Combativeness, Resistance—defense.
7. Destructiveness, Executiveness—force.
8. Alimentiveness, Appetite—hunger.
9. Acquisitiveness, Accumulation.
10. Secretiveness, Policy—management.
11. Cautiousness, Prudence—provision.
12. Approbativeness, Ambition—display.
13. Self-Esteem, Self-respect—dignity.
14. Firmness, Decision—perseverance.
15. Conscientiousness, Justice equity.
16. Hope, Expectation—enterprise.
17. Spirituality, Intuition—faith—credulity.
18. Veneration, Devotion—respect.
19. Benevolence, Kindness—goodness.
20. Constructiveness, Mechanical ingenuity.
21. Ideality, Refinement—taste—purity.
B. Sublimity, Love of grandeur—infinitude.
22. Imitation, Copying—patterning.
23. Mirthfulness, Jocoseness—wit—fun.
24. Individuality, Observation.
25. Form, Recollection of shape.
26. Size, Measuring by the eye.
27. Weight, Balancing—climbing.
28. Color, Judgment of colors.
29. Order, Method—system—arrangement.
30. Calculation, Mental arithmetic.
31. Locality, Recollection of places.
32. Eventuality, Memory of facts.
33. Time, Cognizance of duration.
34. Tune, Sense of harmony and melody.
35. Language, Expression of ideas.
36. Causality, Applying causes to effect. [tion.
37. Comparison, Inductive reasoning—illustra-
C. Human Nature, Perception of motives.
D. Agreeableness, Pleasantness—suavity.

A standard depiction of phrenological organs that the Fowlers used across their materials

ORDER I: AFFECTIVE FACULTIES *(related to emotions and feelings)*

GENUS I: PROPENSITIES *(related to humans as animals and in the physical world)*

SPECIES I: DOMESTIC PROPENSITIES *(related to social, civil, and familial relations)*

1. AMATIVENESS *(love and sexual attraction)*

A. CONJUGALITY *(monogamy and union for life)*

2. PHILOPROGENITIVENESS *(parental love for kids, love for animals)*

3. ADHESIVENESS *(friendship, social connection with others)*

4. INHABITIVENESS *(love of home, patriotism)*

5. CONCENTRATIVENESS *(unity of thought and feeling, continuity of action)*

SPECIES II: SELFISH PROPENSITIES *(related to individual and animal needs and desires)*

E. VITATIVENESS *(love of existence, fear of death)*

6. COMBATIVENESS *(resistance, opposition, defense)*

7. DESTRUCTIVENESS *(force, drive to destroy and cause pain)*

8. ALIMENTIVENESS *(appetite for food and drink)*

9. ACQUISITIVENESS *(drive to possess, personal ownership)*

10. SECRETIVENESS *(tendency to conceal and hide)*

GENUS II: HUMAN, MORAL, AND RELIGIOUS SENTIMENTS *(related to the elevation and ennobling of human beings)*

SPECIES I: SELFISH SENTIMENTS *(related to individual interests and happiness)*

11. CAUTIOUSNESS *(anxiety, apprehension of danger)*

12. APPROBATIVENESS *(regard for character, approval of others)*

13. SELF-ESTEEM *(self-respect, feelings of liberty and independence)*

14. FIRMNESS *(decisiveness, stability, pursuit of purpose)*

SPECIES II: MORAL AND RELIGIOUS SENTIMENTS *(related to morality and religion)*

15. CONSCIENTIOUSNESS *(feeling of justice, accountability, duty)*

16. HOPE *(anticipation of future success, happiness)*

17. MARVELOUSNESS *(wonder, belief in the supernatural)*

18. VENERATION *(worship, respect for sacred things)*

19. BENEVOLENCE *(kindness, sympathy, care for others)*

SPECIES III: SEMI-INTELLECTUAL SENTIMENTS *(related both to feelings and intellectual reasoning)*

20. CONSTRUCTIVENESS *(ability to build, make, invent)*

21. IDEALITY *(imagination, taste, fancy)*

B. SUBLIMITY *(grandeur, beholding what is magnificent)*

22. IMITATION *(ability to pattern and copy)*

23. MIRTHFULNESS *(sense of what is funny, absurd, ridiculous)*

ORDER II: INTELLECTUAL FACULTIES *(related to objects, physical qualities, and abstract relations)*

GENUS I: PERCEPTIVE FACULTIES *(related to the connection between the mind and the external world)*

SPECIES I: EXTERNAL SENSES *(the human senses)*

TOUCH, SIGHT, HEARING, TASTE, SMELL *(unlettered and unnumbered, given disagreements over the place of these faculties among the other faculties)*

SPECIES II: OBSERVING AND KNOWING FACULTIES *(related to the organization and cataloguing of things in the external world)*

24. INDIVIDUALITY *(observing and individualizing things in one's environment)*
25. FORM *(understanding shape and configuration)*
26. SIZE *(apprehension of magnitude, proportion, bulk)*
27. WEIGHT *(apprehension of specific gravity, attraction, force)*
28. COLOR *(ability to see and understand color)*
29. ORDER *(physical arrangement, placement)*
30. CALCULATION *(relations of numbers, ability to calculate)*
31. LOCALITY *(relative position of things and places)*

SPECIES III: SEMI-PERCEPTIVE FACULTIES *(related to the connections among objects and qualities)*

32. EVENTUALITY *(grasp of actions, events, phenomena)*
33. TIME *(apprehension of the passage of time)*
34. TUNE *(melody, tone, musicality)*
35. LANGUAGE *(communication through words, sounds, signs, and symbols)*

GENUS II: REFLECTIVE OR REASONING FACULTIES *(related to high-order thinking, abstract ideas, and human reason)*

36. CAUSALITY *(understanding cause and effect, reasoning deductively)*
37. COMPARISON *(understanding similarities, reasoning inductively)*

C. HUMAN NATURE *(understanding other people and their character and motives)*

D. AGREEABLENESS *(persuading others through charm and disposition)*

Although the Fowlers made phrenological analysis seem easy and could examine a person's head in a matter of minutes, the system was complex and often confounding, which provided ammunition for anti-phrenologists. Some of the organs were small and clustered in tight spaces, like those around the eye. How could an examiner distinguish between Weight and Color, for instance? Other organs rested on the borders of groups that performed different functions.

Sublimity and Cautiousness, for example, were right next to each other, yet they worked in very different directions. Was it really possible to feel the difference between an organ directed at vastness and infinitude and an organ tied to anxiety and potential danger?

On top of these questions was the issue of scale, or the relative size of each organ. When Orson came up with the idea of placing each organ on a scale, he introduced gradations that, he believed, would result in a more accurate picture of a person's character. But he also introduced possible permutations and combinations that made the science exponentially more complex. Although the Fowlers experimented with different scales (at first it was a three-point scale and briefly a twenty-point scale), they ultimately standardized around a seven-point scale:

- 7 = Very large
- 6 = Large
- 5 = Full
- 4 = Average
- 3 = Moderate
- 2 = Small
- 1 = Very small

If that wasn't enough, the Fowlers often indicated half sizes or pluses in their reports. So a person could receive a standard 6 or a 6.5 or a 6+.

But the Fowlers weren't in the business of providing numbers alone. They expertly interpreted those numbers for their sitters, explaining, for instance, the meaning of a large (6) organ combined with an average (4) organ, or a very large (7) organ and a small (2) organ. This was one reason that the Fowlers published lengthy manuals detailing the interplay of possible combinations and ratios. For instance, they noted that if you have large (6) or very large (7) Approbativeness (love of the approval of others) combined with large (6) or very large (7) Adhesiveness (attachment to society), you will be "extremely sensitive to the approbation and the disapprobation, particularly of friends." Yet a large (6) Approbativeness combined with full (5) Firmness and moderate (3) Self-Esteem means that you will generally act in conformity with the wishes of your friends. However, large (6) Approbativeness combined with large (6) Language and Comparison and full (5) Conscientiousness, Veneration, and Causality means that you'll be quick to speak up in the company of friends and to engage others in debate and public meetings. Or take large (6) Approbativeness and layer on very large (7) Ideality, full (5) Causality, and a generally smaller sized head, and you will be "a fashionable dandy, who will devote himself chiefly to dress, etiquette, and tea-table talk."[12]

These were just some of the possible combinations explained in one paragraph specifically about a large Approbativeness. Every other size of every other organ involved similar explanations, which was one reason that the Fowlers' books and manuals were often hundreds of pages in length. Phrenology was not just about bumps; it was a massive cartography of each individual human head.

ON THE ROAD in the fall of 1834, Orson and Lorenzo were succeeding beyond what they could have hoped, which gave them a new idea: If they split up, they could cover more territory, spread phrenology even further, and make more money. So in late fall Orson headed to the East Coast to the nation's biggest cities where phrenology could shape the minds of the masses, while Lorenzo headed west and south, spreading the science to geographic doorways through which the nation was expanding.

Lorenzo's Western sojourn took him to Michigan, Ohio, Indiana, Illinois, and Missouri. He stationed himself in various towns for days, even weeks, and delivered lectures, demonstrated the science, and examined paying customers. In Indiana, a young college student visited him for a reading and an answer to the question of what he should do with his life. Lorenzo felt the young man's head and replied: "You can write poetry, for your Ideality is one of your largest organs, and you have just the temperament adapted to it."

The student let out a chortle. "But I have never written a line in my life!"

"Make the attempt, then," Lorenzo advised him squarely, "for you can succeed."

The young man went away, settled in a shady, tranquil space under a tree, and penned his first verses. Years later, the American people came to know this young college student as William Ross Wallace, a preeminent American poet, whose most famous work was the 1865 ode to motherhood, "The Hand That Rocks the Cradle Is the Hand That Rules the World."[13]

Lorenzo had a similar experience in Cincinnati, although it was yet another test. One day a man dragged his drunken friend to Lorenzo's temporary office, located on Vine Street, for an examination. "Is it possible to make anything of this drunken loafer?" he asked the young phrenologist.

After feeling the drunkard's head, Lorenzo came back with an answer: "Drunk or not, there are the elements of greatness in this brain. With an opportunity for development he can make a mark in the world, for he has ambition, perseverance, fine power of imagination, an excellent eye to judge of proportions, superior natural talent for a first-class mechanic, architect, sculptor, artist, or scientific man."

Hearing the answer, the two men smiled to each other and dropped the ruse. The supposed drunk was actually Hiram Powers, one of the most celebrated American sculptors of the nineteenth century. In fact, Powers had recently spent time at the White House studying the head of Andrew Jackson. In less than a year from the time of Lorenzo's exam, Powers would produce what is considered the most true-to-life representation of Jackson ever created—a bust that still sits in the Smithsonian American Art Museum.[14]

Over the next eighteen months, Lorenzo lectured and examined heads on the Western edge of the nation, which at the time was St. Louis, and he traveled throughout the slaveholding South, which included stops in Lexington, Kentucky; Nashville,

Tennessee; Natchez, Mississippi; and New Orleans, Louisiana. Because he knew that two sets of hands were better than one when it came to phrenology, he hired an assistant. At first the assistant was a Mr. Farnham, but later in the tour he hired Joseph Buchanan, who had studied medicine under Charles Caldwell and, a few years hence, would help launch the science of neurology. Everywhere they went, Lorenzo and his assistant received favorable coverage, with newspapers calling them "truly astonishing" demonstrators whose "accuracy in practical phrenology is so far beyond dispute" because they "have not once erred from the person examined."[15]

As Lorenzo worked his way through the West and South, Orson focused his energies on the Northeast. In early 1835, he toured the Deaf and Dumb Asylum of New York, where students learned to communicate by acting out their ideas; this was the reason, Orson told administrators, they all had Imitation large or very large. After New York it was on to Pittsfield, Massachusetts; Philadelphia, Pennsylvania; and Washington, D.C. Eventually Orson ended up in Baltimore and became part of a public battle over phrenology unlike any he had experienced before. The battle ended in a victory for phrenology, at least according to Orson, and the publication of his first book.[16]

It was the summer of 1835 when Orson arrived in Baltimore, Maryland, and local leaders took a stand against the science. They questioned phrenology, ridiculed Orson, and portrayed his work as nothing less than a blatant money grab. They even showed up to his lectures to rattle him. In early June, Rev. Benjamin Kurtz, editor of the *Lutheran Observer*, attended a lecture and approached Orson after it was done, at which point "a controversy immediately ensued." The heated back-and-forth about the merits of the so-called science finally settled down, at which point Kurtz made Orson an offer. The reverend had two sons, a fifteen-year-old and a thirteen-year-old, and if Orson could examine their heads and provide "a correct description of their intellectual and moral character," Kurtz would yield his opposition to phrenology.[17]

The next day Kurtz, along with "a few other respectable and

literary gentlemen of this city," arrived at Orson's temporary office space with the two teenagers in tow. To keep the test as fair as possible, Kurtz had written out his description of the boys as a kind of experimental control. When the examinations were done, the other men in attendance would compare what Kurtz had written with what Orson explained.

Orson started on the examination and, as he was feeling their heads, described their characters "minutely, fully, and unequivocally." When the exam was done, Kurtz and his associates compared notes and were blown away by what Orson provided—"a most striking and astonishing" description that matched what was on Kurtz's page. They were all now converts to the science. "Nothing but facts, stubborn and irrefutable facts," Kurtz wrote in the pages of the *Lutheran Observer*, would have changed his mind, but that's exactly what Orson provided. "We now verily believe that great injustice has been done to this department of useful study, and to those who, in spite of the taunts and jests of opponents, are zealously pursuing it."[18]

Reading what the minister had experienced, the people of Baltimore visited Orson in droves, posing their own tests of the science. They brought him more unsuspecting children, asked him to examine sitters while wearing a blindfold, and paid him to reexamine earlier sitters so as to compare an old character description with a new one. Feeling somewhat under assault, even as he successfully met these challenges, Orson wrote a letter to the editor of *The Chronicle* in Baltimore to issue a challenge of his own. Given that phrenology has attracted many "violent opposers" in the city, he announced, detractors should "state their objections to" the science in "strong, yet concise terms" so that he could answer them clearly in the pages of the newspaper. This, he hoped, would reach more citizens of Baltimore than lecturing alone ever could. Argument, not ridicule, was the real way to truth.[19]

Writing to the paper the next day, Vindex—a pseudonym for Orson's opponent—indicated that he would be willing to grapple with the young phrenologist in the pages of the paper. But first he needed Orson to explain his take on phrenology, given that "each

Phrenologist has a system of his own." To that end, Vindex posed a series of questions, including, "Is not the skull liable to bony excrescences, and may they not be mistaken for phrenological organs?" Answer these questions, Vindex instructed Orson, and they could then grapple in the pages of *The Chronicle*.

Orson saw his opening. Instead of answering the questions directly, he seized upon Vindex's point that "each Phrenologist has a system of his own." This was patently false, he exclaimed. While there are some differences among phrenological systems, there is "perfect unanimity" on the "fundamental principles." In fact, "There is at the present time greater unanimity among phrenologists than among the teachers of any other doctrine or science within my knowledge."[20]

The battle was launched. Orson and Vindex clashed across numerous issues of the paper, as both men made bold proclamations about the science, questioned the logic of each other's arguments, and homed in on specific words to try to pull the discursive rug from under the opponent. The editor of *The Chronicle* must have loved the exchange, because he began printing longer and longer letters from Orson. While Vindex's letters remained relatively brief, Orson wrote as much as the editor would allow, resulting in lengthy, pro-phrenology columns in the pages of a leading newspaper in one of the nation's most populous cities.

Ultimately Orson and Vindex went back and forth ten times, with Orson producing many more pages than Vindex. Soon after the controversy ended, an editorial in another newspaper, *The Saturday Morning Visitor*, called on Orson to publish his exchange as a stand-alone pamphlet that would preserve the debate for posterity. Orson agreed, and in late summer 1835, *Phrenological Controversy: Answer to Vindex* appeared in local Baltimore bookstores. Combative, uncompromising, and hastily written, yet filled with detail and verve, Orson's first book established the tone and template for the remainder of his decades-long writing career.

WHEN VINDEX GRAPPLED with Orson in the pages of Baltimore's *Chronicle*, he trained his arguments on the scientific merits of deciphering the human mind via the size, shape, and curvature of the head. Vindex, whom Orson believed to be Maxwell McDowall and perhaps other associates, was arguing the issue as a scientist concerned primarily with the status of scientific knowledge. Orson was concerned with the march of science as well, but he understood something that Vindex and many other anti-phrenologists did not appreciate—namely, many Americans were ambivalent about science itself. New knowledge was a good thing, of course, but what really mattered was what people could do with the new knowledge. Vindex treated science as an end; Orson and the mass of followers he attracted treated it as a means to something greater. As Orson wrote in the pamphlet version of his exchange with Vindex, phrenology "leaves every man free to determine his own character, and puts into his hands the power of giving, to a greater or lesser extent, just such a shape to his head as he chooses."[21]

In Orson's *practical* phrenology, science was a means to improvement. By understanding it, practicing it, and cultivating your character, you could become a better human being. Phrenology put the power to improve in your hands. And Orson meant this quite literally. Through phrenology, you had the power to exercise—or refrain from exercising—the organs of your mind, which would increase the size of the corresponding parts of your brain and push out on different parts of your skull. Think of it like lifting weights. If you curl a dumbbell, your bicep will tear then repair itself, thus making the muscle bigger and pushing out on that part of your skin. For phrenologists, the same thing happened with mental organs, which were like the muscles of the mind. Exercise them and your brain will literally deposit new gray matter on the corresponding site, thereby growing that organ and enhancing its mental operations. Grow the organ enough and it will push out on your skull, thinning it at that site and creating a new bump indicative of phrenological development.

To those who thought this notion far-fetched, the Fowlers had plenty of examples of people who exercised different organs, thinned their skulls, and changed the shape of their head. Perhaps their favorite example of this was Benjamin Franklin. Young Franklin had a head "remarkable for observation, memory in general, desire for acquiring knowledge, especially of an *experimental* character, and facility of communication." But old Franklin was different. His head was "all reason and philosophy, rich in *ideas*, full of pithy, sententious proverbs, which are only the condensation of Causality, and always tracing everything up to their causes and laws, but less inclined to observe and remember facts *as such*." Just compare the head of young Franklin to that of old Franklin, the Fowlers insisted, and you can see the change in his forehead, which corresponded to the change in his reasoning powers. His Comparison and Causality started out moderate (3), but they became large (6) or very large (7) because of the way he exercised his mind.[22]

The same head change was open to everyone. In *The Illustrated Self-Instructor in Phrenology and Physiology*—a pamphlet often included with phrenological readings and marked with the sizes of the sitter's organs—the Fowlers concluded the discussion of each organ with advice on how to cultivate or restrain that part of the brain, depending on what the individual needed to do. To cultivate Philoprogenitiveness (love of children), the Fowlers told readers to "play with and make much of children; try to appreciate their loveliness and innocence, and be patient and tender and indulgent toward them; and if you have no own children, adopt some, or provide something to pet and fondle." With enough repeated effort, this spot on the brain would grow and push out on the corresponding spot on the head. However, some people had too much parental love; they smothered and obsessed over kids. In those cases, they needed to restrain the organ: "Set judgment over against affection; rear them intellectually; give yourself less anxiety about them, and if a child dies, by all means turn your mind from that loss by seeking some powerful diversion, and a change of associations, removing clothes and all remembrances, and keep from talking or thinking about them." That was tough but necessary advice at a time in history when parents routinely lost children to disease, accident, and other horrors.[23]

For the Fowlers, improvement was as much a national project as it was a personal one. It was, Orson declared, the "leading characteristic of the nineteenth century," calling people not only to "their own mental cultivation" but to "the intellectual and moral improvement of mankind, especially of the rising generation." Orson even predicted that phrenology would do more "to promote the happiness, virtue, talents, and well-being of man, than has been done by all the other improvements and inventions of this and past ages put together."[24]

It was this kind of foolish, unprovable bluster that made Thomas Sewall seethe with rage at phrenology. The science, he believed, was not only wrong but dangerous, giving people a false understanding of themselves and their responsibility to others. And the Fowlers were leading Americans to ruin and damnation, with dollar signs dancing in their eyes.

A promising young physician with a firm commitment to anatomy and physiology, Sewall graduated from Harvard Medical School in 1812. He moved to Ipswich, Massachusetts, shortly thereafter and established a private practice. But that's when his passion for the human body got a bit out of hand. The local cemetery, Sewall discovered, was a great place to find corpses that could be used in anatomical experiments. In October 1817, he waited for the cover of night, then headed to the graveyard with a shovel in one hand and a small flickering lantern in the other. There he found a freshly dug grave and went to work. After some back-breaking labor, he hit the coffin, cracked it open, hauled the body out, and dragged it back to his office, where he had full rein to probe the brain, heart, guts, muscles, and nerves of his recently deceased neighbor. Over the next six months, Sewall returned to the graveyard at least seven more times, looting corpses that ranged from ten to seventy-nine years old at age of death. Two of the bodies he dug up and dissected were former patients.

But Sewall had not been as careful as he should have been.

A local man who lived near the cemetery reported to authorities strange lights flicking around the graves at ungodly hours. Jumping into action, the underemployed police of the sleepy New England town were aghast to discover eight graves with empty coffins. Then they received a tip that they should talk to the local anatomist about the missing bodies. Inquiring at his office they found the stuff of nightmares: sliced up organs, hastily stashed away limbs, and other remnants of a corpse collection. Sewall was charged with violating an 1815 state law that made grave robbing a felony. Outraged with the charge, he hired the best lawyer in the country, a young Daniel Webster, who was still a decade away from his first term in the U.S. Senate, to defend him. It didn't turn out well for the doctor. Sewall was convicted, fined $800, and forced to leave Ipswich.[25]

Exiled and ashamed—not by what he had done but by how he had been treated—Sewall ended up in Washington, D.C., and worked to rebuild his medical practice. Once again successful in private medicine, he became a professor of anatomy and physiology at Columbia College. At the same time, he grew in his Christian faith and became a respected leader in the Methodist Episcopal Church.

For Sewall, all was going well in Washington, D.C., until those damn phrenologists showed up. Phrenology arrived in the United States long after Sewall's own medical training, and over the years he happily explained to his students that the supposed science was mistaken on almost every count. While he could hold phrenology at bay in the classroom, it was gaining traction in the wider community, and that burned him, especially because the Fowlers were not true medical men. Convinced that phrenology was not only bad science but a threat to Christian morality, Sewall resolved to take a stand against the hucksters who were sowing confusion and reaping a rich reward.

Sewall noticed an ad in the local paper in which Orson called himself a "*practical* Phrenologist" and invited the whole community to a free demonstration at the Unitarian church. Anyone still skeptical of the science after the demonstration, Orson's ad continued, was invited to visit him at Mrs. Mount's Boarding House "to have their own peculiarities of character, disposition, and talents" accurately described. What's more, "Citizens are earnestly invited to test" him in any way they pleased.[26]

Here was Sewall's opening. He attended the demonstration and, as soon as Orson finished speaking, shot his hand into the air to volunteer for a public examination. Orson not only obliged but agreed to examine Sewall while blindfolded. It was a great success for phrenology, at least according to Orson, and even the cantankerous Sewall admitted the accuracy of the character description. But more tests were needed.

The next day Sewall invited Orson to his home to examine his family. Orson complied, examined the whole lot of Sewalls, and described their characters in a way that delighted all of them—except Sewall himself, who remained defiant despite the mounting evidence supporting phrenology. Finally, after verbally jousting over the legitimacy of the science, Orson and Sewall agreed on a test that would, supposedly, satisfy both of them. Sewall would select a number of his friends and write down what he knew of their characters. He would then personally apply the blindfold to Orson to ensure that it was secure. Once blindfolded, Orson would examine Sewall's friends though say nothing of their character aloud. Once the friends left the room, Orson would write down their character. Finally, the friends themselves would compare Sewall's description with Orson's and judge which was the better account.[27]

The test took place at the conclusion of Orson's next lecture. Blindfolded to Sewall's liking, Orson stood behind a chair in the middle of the rostrum and started examining heads. But Sewall had an ace up his sleeve: For an added test he inserted himself into the line of "friends" to be examined. Because Orson had already examined Sewall, there would be two character descriptions to compare, along with the descriptions of the friends that both Orson and Sewell had written out. Once the experiment was complete, the descriptions were read aloud for everyone to hear. As Orson later boasted, "Every one of his friends examined, and *their* friends, attested that mine was the most graphic and correct."

The next day Sewall showed up to Orson's office and still refused to back off on his opposition to phrenology. Frustrated and apoplectic, Orson begged him to explain why, after all these many tests, he continued to oppose phrenology. Sewall finally confessed. Several years earlier, Charles Caldwell, the Kentuckian who had helped bring phrenology to the United States, had announced a series of lectures in Washington, D.C. To support the endeavor, Sewall bought a five-dollar subscription to the series—a common practice at the time for funding up-and-coming lectures—then convinced several other people to pay five dollars to support Caldwell's work. Because of all the subscriptions he had brought in, Sewall figured Caldwell would refund his own five dollars. But apparently Caldwell refused. Angry and feeling somehow cheated, Sewall pledged then and there to work against phrenology for the rest of his life. He proudly told Orson, "I have taken and had my revenge by ridiculing the science of Phrenology."[28]

How true was this story about Sewall and Caldwell? It's unclear. Orson first told it in 1839 in the pages of *The American Phrenological Journal*. He then repeated it decades later in his magnum opus, *Human Science*, published in 1870. Sewall, of course, never corroborated the story, nor did he take the time to deny it. But he did continue to oppose phrenology. In 1837, he published *An Examination of Phrenology*, which consisted of the condemnatory lectures he had delivered for years to medical students at Columbia College.

Perhaps most telling about Sewall's published lectures was that they opposed phrenology based on Sewall's own inaccurate understanding of the brain and the mind. Phrenology was wrong, he maintained, because the brain was not many organs but one organ—"a unit, and the whole organ is concerned in each and every operation of the mind." On this score, Orson was closer to scientific accuracy than Sewell. Phrenology held that the brain was not a whole unit but rather comprised different organs that carried out different functions. The idea that you can apprehend those organs by feeling someone's head was wrong, of course, but Sewall's denunciation was not a matter of defending science from the onslaught of pseudoscience. It was rather an instance of the ever-unfolding development of science itself, which is rife with accuracies, inaccuracies, and everything in between. In the 1830s, Orson and other phrenologists had just as much claim to scientific knowledge as Sewall and other anti-phrenologists.[29]

AFTER SIX MONTHS in Washington, D.C., Orson was ready for a change of scenery. He and Lorenzo, who was fresh off his eighteen-month tour of the West and South, reunited during the summer of 1836 and traveled through several towns in Pennsylvania, including Carlisle, Danville, Bloomfield, and Philadelphia. While in Philadelphia, Orson had the opportunity to visit Moyamensing Prison and to examine the heads of numerous convicts while divining the crime that landed them behind bars. "The correctness of his conclusions was generally corroborated by the admission of the subjects," wrote one newspaper in covering his visit, "as well as by the statements of the keepers themselves, who were acquainted with the crimes with which they were convicted."[30]

Life on the road was challenging, even when one was furthering the science of phrenology. After two years, Orson and Lorenzo needed some stability, or at least a place to call home, and New York City seemed perfect. Moving to the Big Apple, they secured office space at 17 Park Row, opposite Astor House, and a couple doors down from the Park Theatre. It was the entertainment district at the time, full of curious citizens willing to shell out money for a bit of fun. To draw attention to the new office space, Orson and Lorenzo lectured regularly across the city, usually for free. Sometimes they lectured together, other times one stayed back at the office to examine heads while the other worked with an assistant to demonstrate the science for eager audiences. The assistant's name was Samuel Kirkham, and he would later go on to write an influential textbook on English grammar and elocution. Personable and adroit as a phrenologist, Kirkham was even named as a contributing author to Orson and Lorenzo's 1836 book *Phrenology Proved, Illustrated, and Applied*.[31]

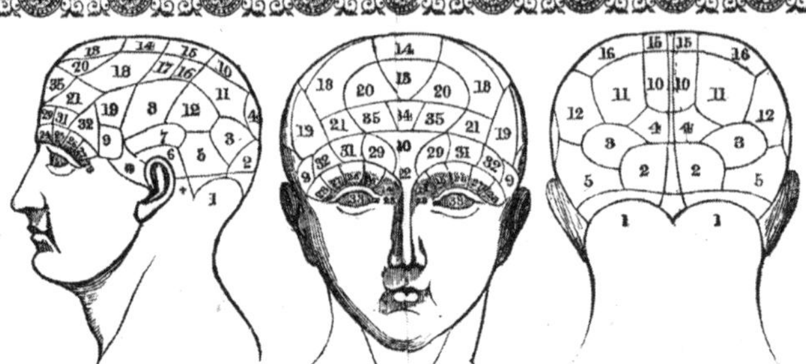

PHRENOLOGY

Practically Applied

TO THE DEVELOPMENT AND DELINEATION OF

LIVING CHARACTERS AND TALENTS,

BY O. S. & L. N. FOWLER AND DR. J. A. BREVOORT,

Practical Phrenologists,

In CLINTON HALL, No. 135 Nassau, or 7 Beekman Street,

Opposite Dr. Spring's Church, New York.

THE citizens of New York, and strangers visiting the city, and also ALL who wish to test the truth of PRACTICAL Phrenology, are respectfully informed, that

THE MESSRS. FOWLER AND BREVOORT, PRACTICAL PHRENOLOGISTS,

Have established their office for the coming year in CLINTON HALL, No. 135 NASSAU, or 7 BEEKMAN STREET; where they pledge themselves to furnish to all who may give them a call, a correct analysis and delineation of their various dispositions, traits of character, peculiarities of talent, capabilities, qualifications, and all other qualities both of intellect and feeling which they or their children may possess, including alike their strong and their weak points of character, their virtues and their failings, with directions how to improve the one to the best advantage, and to correct the other—and all indicated upon a chart, with references to those pages of their book in which these several characteristicks are fully delineated, so that they can be preserved for the future perusal and examination of the individuals examined, or of their friends.

Written descriptions of character can also be obtained.

In connexion with S. Kirkham, author of "Kirkham's Grammar," &c., they have recently published, and offer for sale at their office, a work entitled "PHRENOLOGY, PROVED, ILLUSTRATED, AND APPLIED," accompanied by a CHART, containing 33 cuts, illustrative of the science.

This work consists of between 400 and 500 close duodecimo pages of original matter, and is divided into three parts.

Part first, presents, briefly, the arguments and facts by which the truth of this science is mainly supported.

Part second, extensively describes and analyzes the various faculties of the mind, together with their corresponding organs, in *six states* of development, embracing, also, the modifications of character produced by their relative and combined influence—a part of phrenological science more important than any other, to him who is desirous of learning to *apply its principles to practice*, and, moreover, a part more *imperfectly* treated by other authors than any portion they have handled. Under this head are likewise presented the *phrenological developments and character of several hundred distinguished individuals*, among whom are Washington, Franklin, Jefferson, J. Q. Adams, Jackson, and Van Buren; Webster, Clay, Calhoun, Preston, Duff Green, Wise, Benton, Cass, and R. M. Whitney; Irving, Halleck, Willis, Knapp, Chancellor Kent, Professors Eaton and Duer, Charles King, J. W. Webb and O. Hoffman; Drs. Humphrey, Griffin, Beecher, and Alexander; Rev's. Phinney, Burchard, Ludlow, and Stockton; P. Bennet, Garrison, Frances Wright, Offen, &c. &c.

Part third, presents the *moral bearings* of phrenology, including its *natural theology*, together with *replies to all the important objections* of its opponents.

They have also published a phrenological BUST, which presents improvements over *all others* in the following important particulars:—first, it presents a new *classification* or *grouping* of the organs; secondly, it gives the *correct location* of each of the organs; thirdly, it presents each of the organs *raised*, or as it actually appears in the *living head* when fully developed; and, fourthly, it gives the *names* of the organs written upon each.

They will wait upon individuals, select parties, or families, at their dwellings or elsewhere, either in this city or Brooklyn, and *without extra charge*. They will also instruct either individuals or classes in applying the principles of phrenology to *practice*.

They respectfully invite the publick to call and examine their extensive collection of casts, sculls, &c., in proof and illustration of comparative Phrenology, at their office.

Clinton Hall, New York, April 24th, 1837.

An 1837 flyer advertising the services of Orson and Lorenzo Fowler

When the Fowlers lectured together, audiences often liked to test the brothers against each other. After speaking on the science, one of the Fowlers would leave the auditorium while the other examined an audience member and described the person's character. Then the Fowler who had left the auditorium would return and provide a second description of the audience member's character. If phrenology was a real science, the two descriptions should match, as was the case at a lecture on January 28, 1837, in New York. Facing an audience of over one thousand people, the Fowlers passed what Orson later called the most important public test of phrenology up to that point.

Unbeknownst to the brothers, one audience member that evening was a man named Benjamin Offen, infamous around the city as an "infidel lecturer"—a freethinker and vocal critic of organized religion. At the close of the lecture, the audience nominated Offen for public examination. With Lorenzo out of the auditorium, Orson proceeded with the examination, thoroughly unaware of the identity of the man sitting before him. He discovered that Offen's organ of Veneration was small and his Marvelousness was "almost wholly wanting." This man must be a skeptic, Orson told the audience. But there was more. The man's Eventuality, Language, Mirthfulness, and Imitation were very large, which meant that he was likely a skilled public speaker and debater—a "great reasoner, and would display a great command of words, facts, and arguments," Orson described.

When the exam was done, Lorenzo came into the auditorium and gave his own examination, describing Offen "so perfectly, *on every point*" that the audience grumbled about collusion between the brothers. Once listeners were satisfied that no collusion had taken place, another audience member, Mr. Vail, stood up and revealed that the man just examined was indeed the infidel lecturer Benjamin Offen. Vail then "went minutely into his character, taking it up, point by point, and illustrated most strikingly, and confirmed, each as stated by the phrenologists." Then Offen rose to his feet and praised the descriptions offered by Orson and Lorenzo, confessing his own change of heart about the science. "The remarkable accuracy with which they had delineated the various features" of his character turned the infidel into a believer. Marching proudly forward at the hands of the Fowlers, phrenology had become a gospel message baptizing the nation, even its unbelievers, into a new scientific reality.[32]

CHAPTER FOUR

A CABINET OF CURIOSITIES

ELEVEN MILES AWAY, CHARLOTTE FOWLER HEARD THE exciting news of her brothers' return, jumped on a horse, and galloped through the woods as fast as she could. The journey from Franklin Academy in Prattsburgh, where she had been a student and was now teaching school, to the family home in Cohocton was not an easy one; she had to cross streams, traverse hills, meander through meadows, and endure a jostling saddle. But it was worth it to embrace Orson and Lorenzo, who were visiting the family home amid one of their lecture swings through upstate New York.

The year was 1835, and Charlotte was twenty-one years old. By the time her brothers came home, they had been working as professional phrenologists for over a year, and they were thrilled to share with the family everything they had learned. They even gave their parents and siblings phrenological exams, unlocking the secrets of their minds with the confidence of apostles. Charlotte was not only thrilled to have her older brothers back; she also fell in love with the science they preached. When they departed, they gave her a small book on phrenology. "I read with eager interest and I hungered for it," she later recalled.[1]

Orson and Lorenzo's visit was actually not the first time she heard about the wonders of the new science. Like many Americans, Charlotte learned about it in the wake of Spurzheim's visit to the United States. Horace Fowler, the family patriarch, had a habit of reading to his children from the newspaper in the evenings while the women mended clothing. Hearing her dad read about Spurzheim and phrenology planted a seed for the young woman, and she took it upon herself to learn everything she could. As she later recalled her maturation with the science,

> I began to study heads in 1832, when Dr. Spurzheim was lecturing in Boston . . . I had no teacher but a phrenological bust and a small pamphlet and the heads of my little brothers and sisters to compare with a marked chart. Phrenology at that time was such a novelty that everybody wanted his or her head examined and although I protested that I was only a learner myself, all of my friends and acquaintances begged me to examine their heads.[2]

And examine their heads she did. By all accounts, Charlotte was the go-to phrenological examiner in her part of upstate New York. She even "taught a class of ladies and gentlemen a course of lessons in Phrenology—the first class in that science ever taught in this country." It's unlikely that her class was the first in the country, but it was certainly an early one—and groundbreaking because it featured a woman teaching a class of women and men, something thoroughly taboo at the time.[3]

When her brothers continued on their lecture tour in 1835, Charlotte continued to study and practice phrenology. Her own cranium was well suited to the task. With a round, full forehead, high, arching crown, and smooth-yet-prominent sides, she had a mind for greatness, especially if she could get out of upstate New York. She desperately wanted to join her brothers in promoting the new science. Finally, in 1837, she got her chance; Orson wanted her in Philadelphia to help build something new, something exciting—a phrenological depot unlike anything the world had ever known. She packed up her meager belongings and headed to the City of Brotherly Love.

POPLAR, ELM, AND maple trees dotted the bustling Chestnut Street in Philadelphia. Horses pulling carriages trotted along the cobblestones. Men in top hats and bow ties rushed from shop to shop. Women in flowing white dresses tried to keep their garments out of the road's grime. On each side of the street were tall, stately buildings with classical ornamentation—Corinthian columns, stone carvings, figures in marble. But there were also ordinary buildings, the kind you'd see in any city at the time, where mustachioed men in bowler hats beckoned to passersby to inspect the wares inside their stores.

If you had to name a single street that represented the adolescent nation, you'd be hard pressed to find one better than this. Running from the Schuylkill River on the west side of the city to the Delaware River on the east, Chestnut was home to the U.S. Mint, the University of Pennsylvania, the Jefferson College of Medicine, the Academy of Fine Arts, Independence Square, and Independence Hall. It was also dotted with the storefronts of artisans, craftsmen, and professionals—tailors, milliners, lawyers, painters, carpenters, physicians, jewelers, booksellers, and more. Looking up and down Chestnut, you could see clear visual evidence of American progress, of a nation that had survived its founding and was now embarking on the next phase of its existence.

The Fowlers' storefront at 210 Chestnut Street was a particularly potent symbol of American progress, somehow able to summarize the many elements of Chestnut Street under one roof. It heralded the democratic hope of Independence Square, the dignity of Independence Hall, the higher learning of the University of Pennsylvania, the medical practices of Jefferson College, the artistry of the Academy of Fine Arts, the services of independent physicians, the spirit of booksellers, even the craftsmanship of the carpenters and painters. The skulls that lined its window display whispered of the nation's past while looking toward a proud, progressive, scientific future.

Orson had moved there in the summer of 1837, leaving Lorenzo behind in New York so the Fowlers' growing business had a foothold in two of the nation's top three most-populated cities (the other being Baltimore). They called their Philadelphia outpost the Phrenological Atheneum. In one corner of the store were books and periodicals heralding the new science to the world. In another corner were tools for the practice of cranium examination—charts, diagrams, and reporting documents. Along a side wall were rows of heads—some of them actual human skulls, others casts and busts of famous figures the world over. Visitors could even purchase their own model phrenology head, which at that time was made of plaster and etched with the placement of the phrenological organs. But the centerpiece of the Atheneum was the examination room, where visitors could receive their own reading, understand what they needed to do to improve themselves, then work phrenologically to become stronger, healthier, and smarter. "There is something in the science worthy the attention of the community," wrote the *Philadelphia Botanic Sentinel.* "Visit the Atheneum for yourselves."[4]

Many Philadelphians took the advice. As a writer for *The Pennsylvania Inquirer* summarized, phrenology "is becoming quite popular in this city; and hundreds who a few years since ridiculed it as a humbug, now admit their belief that there is something in it." A big reason for the science's growth, the paper continued, was Orson Fowler, whose demonstrations "had no little effect in exciting inquiry and securing converts. We have heard of some remarkable instances of character described, and with wonderful accuracy, by this gentleman." This was the kind of coverage Orson garnered, largely through a lecturing blitz across the city, where he juggled several series in different venues, "thus lecturing every evening." His standard practice was to deliver a lecture, publicly examine a few heads from the audience, then invite the rest of the audience to receive their own examination at the Atheneum. There they would meet not only Orson but also Charlotte, who managed the store, processed customers, and kept everything running smoothly. Though she never sought the kind of recognition her brothers did, she was the glue that held the whole operation together.[5]

As Charlotte ran the shop, Orson looked for whatever opportunity he could find to promote the science. On one memorable occasion in 1838, he encountered a notorious "atheistic clique" who challenged ministers and churchgoers to debate the existence of God, the possibility of immortality, and the truth of the Bible. Rising to the occasion and always ready to attract an audience, Orson squared off against the group's leader, who had a "forehead of rare height, breadth, and reasoning developments." Though the full content of their debate is unknown, Orson reported that it was an unmitigated victory for him, his science, and his Christian faith. At the end of the debate, the atheist leader turned to his followers and made a bold confession:

> Fellow Atheists, I give it up. You have chosen me your atheistic leader. I resign your leadership commission, for I am an Atheist no longer . . . I always wanted to believe in a God more than others wanted me to, but could accept no dogma, no assertion, nothing without proof . . . This phrenological proof of a God I can understand. Like all of Nature's other truths, it is just as plain as arithmetic. Its inferences fasten to its premises, so as to leave no loophole, no chance for doubt or cavil. It is short, to the point, absolute, and demonstrative.

On top of the bold confession, the former atheist leader began to follow Orson from city to city when he traveled beyond Philadelphia, begging "to hear, over and over again, these doctrines which have opened the windows of my soul to a Divine existence."[6]

Not all of Orson's work in Philadelphia was so dramatic, but much of it was steeped in religion. Early in the nineteenth century, an African American woman named Diana Waters worked as a washerwoman in the city. She was particularly devout, and even more so after she heard the Bible passage "Pray without ceasing" (1 Thessalonians 5:17). Enflamed with the command to worship, she went around the city praying whenever the spirit moved her. She might be carrying a basket of clothes and, seemingly out of nowhere, drop to her knees and address God, bringing "the very heavens down" with her when she prayed. Crowds gathered around her at every turn, even "hundreds of men would stand with heads uncovered" and listen to her wailing petitions. In time, because of extreme overworking of her organ of Veneration, she became deranged, wandering the streets of Philadelphia and "everywhere exhorting all to religious fear and worship, and insisting, no matter how pressing the business in hand, on praying in all the stores and families she visited."[7]

Eventually Diana Waters passed away, and Orson obtained her skull. Knowing her reputation for practically assaulting people with prayer, he carefully studied the underside of her cranium, finding even more proof of his science. Right over the organ of Veneration, on the inside of the skull, was "a white, chalk-colored spot, about the size of a silver dollar." It was as though "that spot had been subject to fire"—like "burnt bone, while all the balance of the inside of her skull was normal." Orson kept Diana's skull for years, until he somehow lost it on the banks of the Susquehanna River during his lecture travels.[8]

At the same time that Orson was creating the Phrenological Atheneum in Philadelphia, Lorenzo was managing the Fowler presence in New York. Early on he partnered with another phrenologist, T. Barlow, who was an expert in making plaster casts of human heads. Together they created the Phrenological Rooms at 286 Broadway. Although the partnership lasted only a year, Lorenzo learned a lot about making casts, a skill that served the Fowlers well for decades to come. With Barlow out of the picture, Lorenzo had to find a new space for his operation, and he found exactly what he was looking for in Clinton Hall.[9]

Clinton Hall had been constructed in the early 1830s for the Mercantile Library Association, which was pitched as an intellectual refuge for middle-class New Yorkers who could, by visiting its reading room and checking out books, eschew the immorality of city life. They could also listen to lectures in a large auditorium and take classes in a variety of subjects. In fact, the most notable thinkers of the era—transcendentalist author Ralph Waldo Emerson, utopian planner Robert Owen, telegraph inventor Samuel Morse, labor activist and minister Orestes Brownson, among many others—delighted and enlightened audiences regularly at Clinton Hall. Orson and Lorenzo themselves lectured there numerous times. While the Mercantile Library Association occupied upper floors of the building, the lower floors consisted of office space that others could lease. Renters at Clinton Hall included medicine companies, fine arts associations, music educators, and even *The New York Herald*.[10]

After Lorenzo found a space to rent, he knew he needed help setting up shop. That's when he reached out to Charlotte about joining him in New York. She readily agreed, especially given her experience running the Philadelphia depot, and she moved in with Lorenzo. For his part, Orson was sorry to see her go, but he was too exhausted to put up much resistance. Not only was he spending endless time promoting and operating the Atheneum, but he was dealing with some challenging family issues as well.

In 1835, Orson had married Eliza Brevoort Chevalier, the first of three women he would marry. Unlike other spouses in the Fowler clan, Eliza would not play a huge role in the family business, save for watching over her and Orson's home and children. The first of those children was Orsena, born in 1838. Another child, named after her aunt Charlotte, followed in 1842. The couple had a third child, a son, in 1845, but he died from illness as a toddler.

Eliza came from a well-to-do New York family. Her father, Elias Brevoort, was a wealthy merchant in New York City. When she met and married Orson, she was a young widow with no children. But she did have a cousin, John Brevoort, who was interested in phrenology and needed a job. Sympathetic, especially to a request from his new wife, Orson invited Brevoort to assist him at lectures and demonstrations. The in-laws worked together in New York, across the East Coast, and as far away as Detroit. When Orson got the idea to establish the Atheneum in Philadelphia, he asked Brevoort to become his partner. There they worked collaboratively to manage the shop and to tour the Northeast spreading the science of phrenology.[11]

Things went well at first, with the in-laws partnering on most aspects of the new business, including phrenological demonstrations. On one notable occasion, Fowler and Brevoort were handed a skull by A. C. Dayton, a physician in New York, and asked to comment on the character of the unnamed person. They felt, analyzed, and discussed the skull, after which they described the person who once occupied it as a man who "would murder for money, . . . sly, mysterious, selfish, . . . destitute of moral principle." Hearing the description, Dayton smiled. The skull in question belonged to "Le Blanc, the murderer of Judge Sayre and family in the village of Morristown, N.J." Destitute of moral principle indeed. In 1833, Antoine le Blanc lived in the Sayre basement for two weeks and did odd jobs around the property. One day he became enraged with Sayre's demands and hacked the judge to pieces with an axe, then bludgeoned Sayre's wife and servant to death with a club. He was quickly tracked down, captured, and hanged. Together, Fowler and Brevoort described Le Blanc's character perfectly—another win for practical phrenology.[12]

Behind the scenes at the Atheneum, however, there was big trouble. Brevoort had recently developed a reputation for the sin of "dissipation"; basically, he was a drunken philanderer. On more than one occasion, Orson had to bail him out of jail. To add insult to injury, Brevoort was bad at managing his money and begged Orson to pay off months of his past-due rent. Orson agreed, took out a loan, and put up his possessions, including the goods of the Phrenological Atheneum, as collateral. When Brevoort couldn't pay, Orson couldn't pay either, and he soon had the sheriff and a collection agent on his trail, apparently at the direction of Brevoort's wife, who "hated me most cordially," Orson later recalled. "So much for intemperance." Fed up with Brevoort, hounded in Philadelphia, Orson found another line of credit, paid off his initial creditor, and shut down the Atheneum. It was time for a fresh start.[13]

Left to his own devices, Brevoort continued a downward slide. Calling himself "Dr. J. Augustus Brevoort," perhaps as a way of distancing himself from his previous reputation, or perhaps simply because it sounded fancy and smart, he continued to lecture on phrenology and to claim a connection with the famous Fowlers. He also started to sing during his lectures, treating his audiences to a demonstration and a ballad about the new science. Orson got wind of Brevoort's singing and thought it the height of foolishness. Yet far worse was Brevoort's ongoing grift. He regularly took money for phrenological publications, then never delivered the goods. In 1843, Orson had to warn supporters that the singing drunk was not his partner, was definitely a swindler, and, worst of all, was tarnishing the good name of phrenology. A year later, Brevoort was arrested for stealing from an old rich widow.[14]

Shuttering the Phrenological Atheneum was ultimately a boon for the Fowlers. When Orson joined Lorenzo and Charlotte in New York, they combined efforts to create something much more influential than they could have if they lived in different cities. At the corner of Nassau and Beekman Streets in Manhattan, they built a cabinet of curiosities that became the headquarters of a scientific, cultural, and business empire.

IF YOU WERE strolling down Nassau Street in the early 1840s, you might happen upon a window display carefully arranged with skulls and heads. Some of them would be of notable Americans—John Quincy Adams and Aaron Burr, for instance. Others would be of criminals—murderers, pirates, and rapists. Still others would be from far-off lands—South America, Africa, and Asia. You might pause at the window display, captivated by those deep black sockets gazing eerily ahead. Then you might look up at the sign above the door: "Fowlers & Wells, Phrenologists & Publishers."

If you went inside the store, you'd likely feel surrounded by the dead, the walls lined with row upon row of skulls peering ominously into a void. Many of these macabre artifacts were once covered by blood and flesh and hair. Perhaps you would imagine the bodies that used to be underneath them, and you'd sense a veritable ghost army watching you as you browse the shelves.

If you walked to the counter in the middle of the store, a

nice-looking bearded man would greet you and invite you to learn more about the skulls. He would almost certainly pull one down from the display, not just showing it to you but putting it in your hands. You'd be holding an actual human skull, and the ghosts of the departed would seem closer than ever. Then the bearded man would start talking to you about what you're holding, and the more he talked, the more you would realize something: This is not some occult oddities shop full of dark, secret, forbidden knowledge; this is a public laboratory designed for learning and enlightenment.

Then the bearded man—he says his name is Lorenzo—would point out something really interesting: Skulls are thicker and thinner at different spots. You'd slide your fingers across the inside cavity and feel exactly what he was talking about. The reason they're thicker and thinner in different places, he would tell you, is because of different levels of activity in the brain that once occupied the skull. When a part of the brain is used frequently, it receives more blood flow, and that blood flow deposits new gray matter to the spots being exercised. That spot then becomes bigger and pushes on the skull, thereby thinning it. So the skull is thinner where the brain is more active and thicker where it is less active. If someone is called thick-headed, you might realize, it's another way of saying they're dumb, and this is the reason why.

Then Lorenzo might hand you another skull and tell you its story. He'd explain that it once belonged to a man named Burly—a drunk, a thief, a philanderer. One day, Burly stole and ate a young calf, and when the sheriff came to arrest him, Burly grabbed a gun and laid the sheriff down.

In your hands, you'd realize in a mixture of horror and excitement, is the skull of a murderer.

Lorenzo would then move your fingers to different parts of the skull—parts he called Alimentiveness, Amativeness, Combativeness, and Destructiveness. You would feel that the skull is indeed thin in these places. Why? Because, Lorenzo would explain, Burly lived a life of drinking, fighting, and screwing. He exercised those parts of his brain all the time.

But there's something even more interesting: Lorenzo would move your fingers to an area on the top of the skull called Veneration. It's thin like the other areas, meaning that it was also active a lot. When Burly was at the gallows for murdering the sheriff, Lorenzo would recount, he dropped like dead weight and the rope around his neck broke, landing him in the dirt, still very much alive. As his executioners fashioned a new noose, Burly began to pray with all his heart—"earnestly engaged in supplicating the Divine blessing." Then the executioner interrupted him and put the new noose around his neck, and Burly was hanged.[15]

You'd understand that in addition to drinking, fighting, and screwing, Burly was a devoutly religious man. And there was no contradiction in his commitments. His damnable vices and his religious devotion were both parts of his constitution.

Then you might set Burly's skull on the counter and wonder about your own brain. Lorenzo would see the interest in your eyes, and he'd ask if you'd like to learn about your constitution. For a nominal fee, he would feel your head and reveal more about who you are than you could ever discover on your own. Fascinated, you would follow him to the back of the store for an examination.

Your experience at the Phrenological Cabinet, located for over a decade in Clinton Hall, was commonplace. Well over one hundred people a day—every day—circulated through the Cabinet to unlock the secrets of skulls and nature. When people were in the city and had some extra time, they made it a point to visit the Fowlers. Or, as the *Paterson Guardian* in New Jersey put it: "When the reader finds himself in the city of New York, with an hour to spare, let him not fail to take a look at Fowler and Wells's collection of skulls and busts." There he will find "all tribes, and kindreds, and nations, and tongues, and peoples—all races, colors, and religions" brought together "in the mute eloquence of a thousand crania."[16]

The idea that the Cabinet contained the "mute eloquence of a thousand crania" was crucial to the Fowlers' success in New York, and they set about cultivating such publicity however they could. One way they cultivated it was by enticing reporters from local

The Fowlers' Phrenological Cabinet, located for many years in New York's stately Clinton Hall

papers to visit and experience the Cabinet, which was exactly what happened with a writer for the *New York Tribune*. He went to their head depot and came away praising the

> large and expensive Cabinet, containing notorious skulls, drawings of particular persons, and busts or casts of distinguished and remarkable individuals of this and other countries . . . The collection is open to the inspection of every person free of charge, and is well worth examining. Phrenology, in the hands of those gentlemen, has assumed a definite and tangible shape; and we advise the young, both male and female—the parent, the teacher, and the man of business—to consult them professionally, if they wish to form suitable associations for life, put their children to proper employments, or obtain the services of those who will serve them faithfully.

That was the heart of practical phrenology—not just learning the science but teaching individuals, families, and professionals to use it as they navigated the marketplace, the domestic sphere, and civil society.[17]

Another way the Fowlers drew attention to the Cabinet was through a parade of advertisements in area papers, grabbing the eyes of readers with an etching of a skull or phrenological bust. "Free admittance to the Fowler's Phrenological Cabinet," proclaimed an ad in *The New York Herald*, "where may be obtained both Phrenological and Physiological Books. Also verbal and written descriptions of character, both day and evening." "PHRENOLOGICAL EXAMINATIONS," shouted an ad in all caps in the *New York Sunday Dispatch*, "giving directions to the most suitable occupations, with verbal or written descriptions of character given whenever desired." The ad appeared next to ads for "McAlister's All-Healing Ointment," "Dr. Townsend's Sarsaparilla," "Count Rumsford's Cookstove," and "Christy's Minstrels," which featured white performers in blackface singing, dancing, and ridiculing African Americans in front of "crowded and highly respectable audiences."[18]

Still another way the Fowlers attracted attention to the Cabinet was via free public lectures on phrenology. Some of these lectures happened in Clinton Hall in the auditorium right above the Cabinet, which listeners could visit for their own examination immediately after the talk. But they also rented lecture rooms at venues around the city, sinking upfront costs into securing the space, allowing people to attend for free, and ultimately using the occasion to sell goods and services. Lorenzo, for instance, spoke regularly at St. Luke's Episcopal Church at the corner of Hudson and Grove Streets in Greenwich Village. He covered a range of topics in these lectures, including the "Improvement of the Young and to Self-Culture," after which "Heads selected from the audience" were examined.[19]

The Fowlers knew that if they could only entice people to visit the Phrenological Cabinet they could make their case for the science of improvement. Lectures and advertisements were important, but the truth of phrenology was tactile. Touching skulls, feeling heads, holding humanity in your hands—that was how phrenology would make its biggest impact. So enter the Cabinet for free, the Fowlers declared to the world, and caress human nature; then stay for an exam of your own. For one dollar, you received an oral description of your character, along with a chart indicating the sizes of your various phrenological organs. For three dollars, you received the same chart but with a written description of your character, which you could cherish as a scientific marker of your inner life.

When the Fowlers first set up shop in Clinton Hall, their office was located at 135 Nassau Street. Then in 1842 a larger space opened up next door at 131 Nassau. In 1849, they were able to expand again, securing the office at 129 Nassau and keeping what was in 131 Nassau. Whatever size of space they rented, the Fowlers lined their walls so the full range of humanity was on display and available to handle and hold. You could feel the contours of famous heads, such as the intellectual, statesmanlike cranium of John Quincy Adams, who was "remarkable for memory, firmness, and force." You could grasp the genius of George Combe, Scotland's leading phrenologist, whose organs were all large, except Constructiveness, which was moderate, and Calculation, which was small. (He was no good at math.) Even better: You could inspect the divine head of Franz Gall, the discoverer of phrenology, whose brain was large and temperament vigorous. Alongside these and many other famous heads were skulls representing the diverse mysteries of humanity. There was the head of a Cherokee chief, a Choctaw chief, and a "Chinaman." There were several heads of unnamed "idiots" and insane people. Then there were people known for dark deeds—murderers catalogued only by their last names (Johnson, Kohl, and Lacenair, to name a few), and several pirates, including the prolific poisoner Jacque Alexander Tardy, whose head was "low" and "villainous."[20]

An especially prized skull in the Fowlers' collection was that of Patty Cannon, whose brutal crimes stood out in pre–Civil War America for their utter depravity. Cannon owned a tavern and boardinghouse on the border between Maryland and Delaware and oversaw a "gang of ruffians," including several family members, who carried out whatever orders she barked. The Cannon gang started their legacy of brutality with theft and murder. At times, slave traders would come to the tavern and boardinghouse with a thick stack of cash in their pockets to buy human beings. Realizing as much, Cannon entertained the travelers for a while, then, when they departed, instructed her gang to follow them down the road, shoot them dead, and take their money. Other times, when road ambushes were too risky, Cannon simply sneaked up behind the visitors and stabbed them in the back. Her followers then hauled the bloody corpses to the yard and buried them in shallow graves.[21]

After years of robbing and murdering, Cannon realized she was overlooking the chance for real money. Instead of robbing slave traders, why not just trade her own slaves? Because Delaware and Maryland, along with the nearby city of Philadelphia, had sizable populations of free African Americans, Cannon concluded that she could capture free people of color and runaway slaves and sell them to buyers in the South. So she had the upstairs rooms and attic of her tavern converted to prisonlike chambers straight out of a nightmare. Soon the establishment was filled with imprisoned Black people who were beaten, brutalized, and awaiting transport to the South.

Known for her fits of rage, Cannon would sometimes snap at her prisoners and kill them on the spot. She was especially irritated with Black women and children, who she feared would make noise, cause trouble, and expose her operation. If a child cried or annoyed her, Cannon might simply bash the child's head in. Once, when a mixed-race child was born to one of her prisoners and she surmised that a gang member was the father, she killed the kid on the spot with a long metal pike. On another occasion, Cannon became so annoyed with a five-year-old child of one of her captives that she grabbed her favorite metal pike and beat him bloody. When the child did not die but continued to scream, she shoved his face into the attic fire and cooked him to death.

In 1829, police caught up with members of the Cannon gang, and they quickly flipped on Patty and revealed the horrors they had witnessed (and perpetrated) over the years. They also identified where on the property police would find bodies and bones. That was enough to send authorities to the tavern and boardinghouse, where they arrested Cannon and found human remains scattered across her back field. She was taken to a jail in Georgetown, Delaware, where she was indicted on several counts of murder. Unwilling to stand trial, Cannon poisoned herself with arsenic while behind bars, and officials buried her in the jailhouse yard (a relatively common practice at the time).

About ten years later, Lorenzo traveled to Delaware to lecture on phrenology, and local authorities decided to test the new science. They exhumed Cannon's body, removed the skull, and presented it to Lorenzo, giving no indication of its provenance. Feeling the skull, moving his fingers across the lumps and spans of the head, the phrenologist was awestruck. It was the skull of an adult woman, he knew, and she was smart, but her moral sentiments "were almost entirely deficient." Lorenzo continued: The woman was "selfish, sensual, deceitful, and cruel to the lowest degree, shrewd, artful, sagacious in laying plans . . . She would be almost indifferent to the principles of justice as well as to human suffering; had a violent temper, great energy, tact, management, and force of character." As a result, she could "exert an extensive influence over the lower order of minds." Hearing Lorenzo's account, the people of Georgetown were gobsmacked by "the remarkable correctness" of the reading. As a show of thanks, they gave him the skull for his collection.[22]

ORSON AND LORENZO began collecting skulls almost as soon as they started their career as practical phrenologists. While it's unclear when they got their first skull, Orson was comfortable enough by 1837 to label what he had amassed as a "collection." Growth of the collection continued in spades. By the end of the 1840s, he reported spending over $10,000 (around $400,000 in today's money) gathering artifacts for the Phrenological Cabinet. Although the exact size of their collection is difficult to determine, one account from 1861 estimated total artifacts at the Cabinet at 4,000, which included 300 human skulls, 200 animal skulls, 500 casts or busts, and 3,000 portraits or drawings. Although most of the Fowlers' specimens were depictions on paper, 300 human skulls and 500 casts and busts, not to mention 200 animal skulls, made for a special assemblage of crania.[23]

To be sure, their collection was just one of many such collections around the world—a fact that raises an important question: Where did all of these skulls come from? The short answer: everywhere. They came from wherever there were dead people. For the Fowlers, this meant regular visits to hospitals, prisons, asylums, and morgues, where they made nice with staff members and got their hands on the bones of the recently departed.

That was the case with the execution of James Edger on May 9, 1845. Edger was an Irish immigrant new to the country who one evening got drunk, ranted about his wife, and ended up stabbing an associate in a tavern. He was tried, convicted, and sentenced to death. Though adamantly opposed to capital punishment, the Fowlers saw an opportunity to expand their collection, so they sent an associate to the execution to persuade the attending physician, who happened to be a proud phrenologist, to hand over Edger's skull. Something similar happened with a sixty-year-old woman who "died in the poorhouse, was taken to the dissecting-room, and found to be a virgin." After persuading the undertaker to hand over her skull, the Fowlers found the reason for her lifelong virginity: Her organ of Amativeness was incredibly small, leading to acute "sexual indifference."[24]

In addition to collecting heads from the recently departed, the Fowlers regularly received human skulls in the mail. In 1846, Joseph Stayman, a dedicated phrenologist, witnessed the Battle of Palo Alto, the first major battle of the Mexican–American War. When the fighting concluded, he combed the killing field for choice human remains, coming away with many skulls that bore "the marks of either the saber or the bullet." He sent them to the Fowlers to become part of the master record of human heads at the Phrenological Cabinet. That same year, the Fowlers received a "fine collection of Chinese skulls" from "Capt. Gardner," whom they called "a thorough-going phrenologist, and sails with a phrenological crew."[25]

For their part, the Fowlers actively encouraged travelers to contribute to the cause of phrenology. "Those who have it in their power to add rare animal and human skulls to our collection," they wrote to supporters, "will lay both us and the phrenological and scientific world under infinite obligations of gratitude, and most effectively advance this science of the organic and mental interrelations." In 1853, the Fowlers published a hearty thanks to Washington Bates, who obtained several skulls while visiting the Sandwich Islands (which later became Hawaii), including the skull of a prominent Sandwich Island chief, which was plundered from the catacombs of Waimea. They used the opportunity to encourage other readers to follow Bates's lead: "By this presentation, he has rescued it from oblivion, and placed it where it will be inspected by tens of thousands, and for generations to come, and thus done great good . . . We earnestly invite sea captains, foreign travelers, huntsmen, and all others who come across the skulls of rare animals, or of foreign tribes of men, to forward them to our collection . . . on account of the good they will do mankind."[26]

When Orson and Lorenzo traveled to lecture on phrenology, they often received skulls as gifts, particularly in the wake of a successful "test" of their science. In Chester County, Pennsylvania, a physician approached the Fowlers after one of their lectures and presented the skull of a formerly respectable woman who, "despite the entreaties of her friends, abandoned herself to the unrestrained indulgence" of her base instincts. Why would this formerly respectable woman descend into a life of cheap, drunken sex? Feeling the cranium, the Fowlers knew why: The sections over Alimentiveness and Amativeness—the parts of the brain responsible for food, drink, and sex—were "thin as paper, and transparent," showing just how much mental activity had gone to supporting her loose lifestyle.[27]

When Lorenzo toured the South in the mid-1830s to speak about phrenology, he received slave skulls as gifts. One such skull came with an incredible story about the enslaved person who once occupied it. Apparently the man was "notorious for his propensity to steal" and had been whipped mercilessly for thievery. Despite the whippings, he continued to steal (or so the story went). Enraged, the slave master "seized an axe and struck it through his skull into his brain." The man miraculously survived the blow, and much to everyone's horror, he continued to steal. Years later, after he had passed away, a local physician got hold of the skull and presented it to Lorenzo, who discovered it was "remarkably thin and transparent at Acquisition and Secretion." Even something as horrific as an axe blow to another part of the brain could not curb his enthusiasm for more stuff.[28]

The Fowlers and their compatriots discussed slave skulls with a matter-of-fact bluntness that seems horrifying today. Like many scientists at the time, they did not consider the provenance of these skulls to be a problem. Today, there is a widespread effort among academics, museums, and scientific investigators to return kidnapped skulls to the communities from which they were looted. Phrenology was a direct beneficiary of the unethical practices that propelled the skull trade for decades. The Fowlers and countless other scientists at the time saw the pilfering of human remains as not only ethical but a service to humanity; they were rescuing the skulls from oblivion.

For the Fowlers and other participants in the skull trade, the ideal collection was one of variety, which meant doing whatever was needed to gain a diverse array of heads. Not only did they hoard slave skulls, but they also gathered as many American Indian skulls as they could, most often by accepting gifts. For instance, several backers sent in specimens of "Flat-Head Indians," who got their name because tribal elders deliberately flattened the heads of young children as a marker of social status. In addition, Orson acquired the skull of Red Sleeves, an Apache chief who was tortured and executed by U.S. soldiers in 1863. D. B. Sturgeon, a physician at the fort where Red Sleeves was murdered, approached the body "a few minutes after he was shot" and quickly "prepared the skull" for the famous phrenologist. Orson considered it "one of the best contributions to phrenological science possible to be made . . . I have never seen anything even in any Indian head which bears any comparison with his as to Cunning, Destruction, or the perceptives."[29]

Collecting the skull of a recently tortured and executed Native American leader was problematic, to say the least; but perhaps even worse was the extent to which phrenologists relied on grave robbing to build their specimen collections. The scientific hunger for skulls that spanned America and Europe in the first half of the nineteenth century created unfortunate incentives for shady characters to dig up the bones of people at rest. There is no clear evidence that the Fowlers directly participated in grave robbing, but they certainly profited from the practice and expanded their collection via ill-gotten specimens.

When the famous Scottish phrenologist George Combe came to the United States in 1838, he took part in several grave-robbing expeditions, including in the Boston area. Combe did not gift any of these stolen treasures to the Fowlers, but he did allow the siblings to make replicas. In fact, he let Orson and Lorenzo copy all of his "splendid collection," as the Fowlers called it. Apparently, Combe later regretted letting the Fowlers duplicate his skulls; they seemed to have damaged some of the specimens, leaving a bad taste for American phrenology in his mouth.[30]

Nevertheless, the Fowlers earned a sizable reputation for their skills at head replication. An article in *The American Medical Almanac*, for instance, praised the "important improvements in the art of taking casts and busts" made by the Fowlers. "They are now able to take with ease and safety, facsimiles of the living head, as correct almost as life. These gentlemen have already taken the busts of more than two hundred of our leading and distinguished men." These leading and distinguished men included such politicians as Thomas Hart Benton, Aaron Burr, Henry Clay, and James K. Polk; such writers as William Cullen Bryant and Horace Greeley; such ministers as William Ellery Channing; such reformers as Horace Mann; and such inventors as Samuel Morse. Orson was also able to make a cast—from life—of the head of Harrewaukay, a New Zealand chief who was also, as the Fowlers delighted to point out, a cannibal. The chief was "on display" at P. T. Barnum's American Museum, right up the street from the Fowlers' Cabinet, and in 1844 Orson convinced him to sit for the casting process, which he did "very cheerfully, and even made some fun over it." The cannibal, whose history involved "some thrilling stories of barbarism," was pretty affable to the whole thing. Orson was ecstatic with the resulting specimen, which revealed a head so extreme that it furnished "a stronger proof, for or against the science, than a thousand other heads . . . This single specimen of the barbarous race will exceed in value a score of subscriptions to the entire volume of the Journal."[31]

Making casts of famous heads was an expensive endeavor. The Fowlers typically paid sitters $30 for the ordeal (around $1,200 today). Those payments added up, and in the early 1840s Orson reported that he ran into serious financial difficulties and had to pause his collecting efforts. Yet he knew that because of his work, "future ages will read, in the language of demonstration, the true characters of many the distinguished men of our age." When his finances improved, he began acquiring heads once again, plowing his profits back into the science he served. "If anyone is curious to know what O.S. Fowler had done with all the money he has made," Orson wrote in the third person, his collection was the answer. "And let the public remember that he is no miser, no spendthrift, but returns to Phrenology all he receives."[32]

SOON AFTER OPENING, the Phrenological Cabinet became the hub of phrenology for the entire nation. Not only was it a leading tourist attraction and the center of operations for the increasingly famous Fowler siblings, but it started unifying the nation's many local phrenological societies under the banner of practical phrenology. In 1841, Orson estimated that between forty and fifty local societies existed in the United States. By the end of the decade, additional groups had sprung up in Lowell, Massachusetts; Concord, New Hampshire; Paterson, New Jersey; Baltimore, Maryland; Ripley, Cazenovia, Watertown, Auburn, and North Lancaster, New York; Berlinville, Greenville, Chagrin Falls, Republic, and Lykens, Ohio; Janesville, Wisconsin; Auburn, Alabama; and many other places. "We say, without fear of contradiction," Orson reflected in the late 1840s, "that Phrenology never before stood as high in public estimation as it now stands . . . It is *true*, and truth *must* prevail—*is actually prevailing*."[33]

To spur on the work of these societies, the Fowlers sold replicas of their skulls, casts, and busts. For "the very low price of twenty-five dollars," they would send out forty of their "best specimens." This included replicas of John Quincy Adams, Aaron Burr, Henry Clay, Napoleon, Voltaire, Harrewaukay, a German murderer named Gottfried, an unnamed idiot from Manchester, an "unnamable savage," and a "Good negro." Buyers would also receive a copy of their world-renowned phrenological bust, "designed especially for learners." This starter pack of scientific artifacts gave the Fowlers a sizable mailing list of supporters across the country who would advance their initiatives, buy their stuff, and echo their message of improvement. At one point, a group in Canada was hoping to start a new society and convinced the Fowlers to replicate their entire collection—hundreds more than the forty specimens in the starter pack. It was a good deal for the siblings: Not only could they turn a small profit, but they could push practical phrenology internationally.[34]

In all these efforts they were earnest. While they ran a business and made money, they believed wholeheartedly in their mission to improve individuals, the nation, and the world. As they explained in the late 1840s,

> The human race has been groaning for centuries under a false system of education, intellectual, physical and social; contending religious factions have embittered life, retarded the expansion of a warm and general benevolence, and in some unhappy instances have deluged the world with blood; criminal jurisprudence and legislation have been sadly at fault; and laws and treatment of insanity have been a sealed book, until Phrenology threw a blaze of illumination on the dark picture. Is it not time that a mighty effort were made to reduce mental science to a practical system, and make it so generally known in every hamlet and home in our land, that every mind

> shall be enlightened and blessed; and laws and theology predicated in harmony with the nature which Deity has enstamped on his creature man?

This divine calling reverberated through the skulls of the Phrenological Cabinet and the gospel message the Fowlers spoke to the entire world. With every convert they won, with every "hit" the science achieved, with every skull they collected, with every local society they helped to create, the Fowlers pushed their phrenological business ever closer to being an empire.[35]

CHAPTER FIVE

SCIENCE FOR THE PEOPLE

DAVID MEREDITH REESE WAS SKEPTICAL OF ALMOST ALL gospels, except for the one at the start of the New Testament. So when he heard about some asinine "science" working its way across the United States that would supposedly save humanity, he knew he had to speak up.

Born in Maryland in 1800, Reese was acerbic and crotchety even at age nineteen when he graduated with a medical degree from the University of Maryland. Quickly rising through the ranks of his profession, he became physician-in-chief at Bellevue Hospital in New York City, a post that afforded him a front-row seat to all kinds of faddish foolishness creeping into modern society. In 1838, he published *Humbugs of New York*, a warning to all righteous citizens about the intellectual landmines that surrounded them. The humbugs he pointed out were all extreme positions—"ultra-temperance," "ultra-protestantism," "ultra-sectarianism," and "ultra-abolitionism"—which, he believed, left behind the wisdom of the founders and their "clearly right" plan for the gradual emancipation of the slaves. In the realm of medicine, which he was particularly qualified to pronounce upon, extreme positions included animal magnetism, homeopathy, and "quackery in general."[1]

Reese saved special vitriol for phrenology. Calling it the most "prevalent and prevailing humbug" of the day, he ridiculed not only the science and those who touted it but the rubes who paid money for a phrenological reading. Based on "intrinsic absurdity and nonsense," phrenology had become a fad because its proponents were able "to enlist public sympathy and impose upon popular credulity" in the name of the almighty dollar. The idea that the soft, mushy brain could push out on the skull, create bumps, and change the shape of a person's head was utter lunacy. Yet phrenologists told a good story and gave people hope—or, rather, false hope. They felt "the bumps of some blockhead," lamented Reese, and poured "into his ears the discovery that he has 'organs' qualifying him for great literary and moral elevation." Understanding that "flattery is a correct coin among their dupes," phrenologists like the Fowlers made a killing from "illiterate, stupid, indolent, and conceited" knaves who lined up for an examination.[2]

This kind of attack irked Orson, Lorenzo, and Charlotte to no end. To Reese, phrenologists were grifters out to take the money of the ignorant masses. To the Fowlers, phrenology revealed the pathway of hope and progress for all Americans. Though they charged money for their work, they didn't spread their gospel to make money; they made money in order to spread their gospel further.

To Orson in particular, Reese's attack represented a distinct threat—one more dangerous than the attack Thomas Sewall had launched a couple years prior. Sewall had tried to convince scientists and medical students of phrenology's folly; Reese wanted to sound the alarm for ordinary Americans. *Humbugs of New York* was written for the very people Orson was trying to unite under the banner of practical phrenology. Even worse, the book lumped the science together with other extremes as a way of undercutting progressive social change writ large. Not just phrenology, then—progress itself was under attack.

The Fowlers knew they needed to respond to Reese and similar critics in an organized, systematic way. That meant not just playing defense but going on offense. It was time for the Fowlers to spread their message beyond the lecture stage and examinations rooms and to put it in the hands of the American people themselves.

SPREADING THE MESSAGE of phrenology further and farther than ever before meant getting into publishing, and the 1830s was the perfect time to do so. Recent technological advancements enabled small firms to print material better, faster, and cheaper than at any time in the past. The most significant such advancement was the steam press, otherwise known as the power press. Prior to 1820, almost all printing was done by hand. Skilled craftsmen—usually two per machine—carefully laid down paper and rolled sticky ink over it on a large, lumbering press. In a highly efficient hour, they could produce around 250 singled-sided sheets of printed material. But due to advancements in Europe in the 1810s and in Boston in the 1820s, printers gained the ability to produce well over 1,000 sheets of quality material per hour. This brought the cost of printing down, led to the creation of more printing machines, and greatly expanded the output of printed pages.[3]

Many Americans saw the power press as a boon for democracy, enabling "a cheap product for a wide audience." That's exactly how Orson saw it, too, even before he left Philadelphia to join his siblings in New York City. Dreams of reaching the masses with news of phrenology started swirling in his head right around the time he established the Phrenological Atheneum. He had looked across the Atlantic to phrenologists in Edinburgh, who published *The Phrenological Journal and Miscellany* and spread the science across Europe. Why not follow their lead? Why not start a journal for the Fowler brand of practical phrenology? Thus was born *The American Phrenological Journal and Miscellany*, which had a rather uninspired name but which underscored the American character of what he hoped to publish.[4]

The journal, he resolved, would be his defensive and offensive weapon in the fight for practical phrenology. "There does not exist on the American continent," Orson wrote in the prospectus he crafted to attract subscribers, "a single periodical whose object is to advocate [phrenology's] truths, repel the attacks made upon it, or answer the enquiries which even candid persons are disposed to make concerning it." His journal would thus publish "the very numerous facts, confirmatory and illustrative of the truth of phrenology" and show "the true bearings of this science on education (physical, intellectual, and moral), on theology, and on mental and moral philosophy." Moreover, the journal would convey the facts of phrenology in a "*decidedly evangelical*" tone, aimed at nothing less than "the establishment of Truth." With such a righteous aim, Orson pledged that profits from a large subscriber base would not go to lining his pockets but to "the enlargement and improvement of the work, without an increase of expense to the subscribers. More frequent illustrations and embellishments will, in that case, be inserted, and the attractions of the work be thus multiplied."[5]

Orson printed 30,000 copies of this prospectus and mailed them to newspapers, learned societies, universities, and influential thinkers across the country. The effort netted about 1,500 subscribers—not a huge number, but also not horrible for a new periodical about a burgeoning science.[6]

When the first issue of *The American Phrenological Journal and Miscellany* appeared in October 1838, Orson leaned heavily into the democratic ethos he hoped to tap into and advance. The journal's goal, he announced, was to make phrenology "a science *for the people*." Because ordinary Americans "are capable of understanding its principles, of applying them, in more than their great outlines, and of tracing them out, to their results and dependencies," his journal would center on "*practical* phrenology," designed for a nation with "a greater variety of character and of talents than in any other single nation upon earth."[7]

Despite this lofty goal, the first volumes of the journal did not go exactly as Orson had hoped. Because he was already stretched thin promoting phrenology across the country, he had hired a young medical student named Nathan Allen to serve as the journal's editor. This proved to be a mistake. Uninspired by Orson's lofty ideals and democratic appeals, Allen made the journal primarily a journal for scientists, much like the Edinburgh *Phrenological Journal* was. But that was a failing strategy in America, and the dwindling subscriber base proved as much. Although Orson's early promotional efforts had netted 1,500 subscribers, only about half that number ponied up money for volume two. At the close of volume three, only around 400 people paid for *The American Phrenological Journal*. That was not even enough to cover the salary of the editor.[8]

Orson fired Allen, then moved to New York to partner with Lorenzo and Charlotte at the Phrenological Cabinet. Now partners in the same city, the siblings faced perhaps the biggest crossroads of their growing business: What should they do with the journal? It was on life support, hemorrhaging money and subscribers with each passing year. Was it worth continuing?

They knew that practical phrenology was a marketable science. In 1840, Lorenzo had come up with his own idea for a publication, *The Phrenological Almanac*, and it was flying off store shelves. The periodical tapped into the almanac craze of antebellum America, publishing calculations about the sun, moon, stars, tides, and seasons as well as articles about the history of phrenology, the phrenological organs, the heads of murderers and idiots, and the activities of the Fowler siblings. Released annually, priced at six cents, and coming in at around fifty pages, *The Phrenological Almanac* sold in the tens of thousands of copies. Why bother, Lorenzo wondered, with continuing a journal that sold only in the hundreds?[9]

But Orson stood his ground. Despite Lorenzo's doubts, he had faith in the journal's future and turned to Charlotte to cast the tie-breaking vote. Thinking the matter over, thinking about what they were building, she turned to Orson and issued her decision: "Brother, you *can* write what people will be delighted and profited to read."[10]

Assuming the role of editor with the start of volume four, Orson pledged to infuse his own experience as a leading phrenologist into the pages of the journal. "I design," he wrote to readers about the new direction, "to render it highly *practical*, and to adapt to the *million*" the lessons he had learned in his years as a phrenologist. This meant telling stories of his travels and interactions with all types of people, and providing high-quality illustrations of individuals, skulls, and bodies. It meant obtaining unique specimens and analyzing them specifically for ordinary Americans, which required some stylistic changes: "I shall endeavor to render my *style* clear and forcible, rather than finished or elegant . . . A prevailing error of modern writers is their substituting beauty of style for strength of thought—there laboring to produce a finished composition to the comparative neglect of their *subject matter.*" None of that would happen with Orson as editor. "From my extensive intercourse with the mass of American minds, I believe them to be thoroughly *practical.* They do not require deep, profound, labored, learned, or lengthy essays, but something short, plain, to the point, and that they can understand at a glance."[11]

True to his word, Orson made the journal increasingly practical and American. He published phrenological readings of notable figures complete with attractive illustrations. Profiles of American presidents—including George Washington, John Adams, Thomas Jefferson, and Andrew Jackson—appeared in the pages. So did profiles of famous politicians, including Henry Clay, John C. Calhoun, and Daniel Webster. There were also profiles of the poet Edgar Allan Poe, the newspaper editor Horace Greeley, the inventor Samuel Morse, and the minister Henry Ward Beecher. Some, but not all, of these famous figures had received hands-on phrenological examinations. The ones who hadn't, including the dead ones, had their characters described through a bust, painting, illustration, or photograph. The Fowlers had become adept—or so they claimed—at deciphering phrenological characters by simply looking at an image of a person.

Orson also printed in the journal letters from phrenologists across the United States, updating readers on the progress of the movement and depicting phrenology as a science taking the world by storm. One particularly intriguing letter arrived at the Phrenological Cabinet in 1846, along with a mysterious box. Inside the box were casts of skulls—reproductions made in plaster—and a note that challenged Orson to provide phrenological readings of the casts without any information as to the origins of the heads. Then, the note said, Orson had to publish his findings in the pages of his journal so the world could read along as the experiment unfolded. Once the findings were published, the anonymous sender would reveal the

ARTICLE XXV.

THE PHYSIOLOGY AND PHRENOLOGY OF HENRY WARD BEECHER, WITH A LIKENESS.

No. 8. Henry Ward Beecher.

All great men have their strong points. On these their greatness depends. Take these away, and they become Samsons shorn—weak like other men.

Henry Ward Beecher, though till recently unknown out of his limited western sphere, is deservedly rising into favorable notice more rapidly than any other man in this country, consequent on the corresponding strength of these points of his character, which are mainly four.

An article in The American Phrenological Journal *detailing the phrenological character of Henry Ward Beecher, famous American preacher and close friend of the Fowlers*

origins of the heads, along with a history of each person, thereby verifying or falsifying what Orson had said. It was to be an "indubitable test of the truth of this science."[12]

Always ready to demonstrate the truth, Orson readily agreed to the challenge. In June 1846, *The American Phrenological Journal* published his report, along with engravings that he commissioned so readers could visualize what he was analyzing. One of the casts, Orson explained, revealed a head large over Firmness but sloped down at Benevolence and low at the crown. The head was also low at Conscientiousness but large at Destructiveness and Secretiveness.

"No phrenologist," wrote Orson, would hesitate to deem the owner of the skull "a most depraved and desperate personage." He was likely a villain, a murderer, a thief, and prone to "gross sensuality" with women of dubious character.[13]

Hard to believe, but the other skull had "still worse" organization. The owner of this head was "a captain in sin"—dastardly himself and able to lead others into wickedness. His intellect was quite strong but turned toward evil. "His immense basilar region undoubtedly directed that intellect upon the commission of crime," Orson wrote. This meant that he and his associates obtained money however they could, did not hesitate to murder along the way, and felt no remorse for their cruelty. At the same time, the cast revealed signs of good speaking abilities, which meant that the owner could tell great stories about his escapades and influence others. He was the "head devil" of some band of unrepentant raiders.[14]

Despite the evil inclinations of these heads, Orson was having fun with the test—so much so that he brought Lorenzo in on the action. To make the experiment double-blind—or at least as double-blind as the Fowlers figured an experiment could be—Orson asked Lorenzo to examine the casts on his own and to write a report. Lorenzo would not see what Orson had written, and Orson would not see what Lorenzo had written prior to setting the report for publication.

Lorenzo's reading of the skulls appeared in the July 1846 issue of the journal, "without either of us knowing," Orson editorialized, "what the other had written, until both were completed, thus furnishing the most perfect and thorough test possible of phrenological science." Lorenzo found mostly the same attributes as Orson, although he expressed his conclusions in more moderate language. He found the first skull to indicate a man who "showed cruelty and revenge without much restraint." The man was fond of women and had a large appetite for goods of the world. He also had a secretive, selfish nature that led him to desperate acts of thievery. Lorenzo found the second skull to indicate a man who was tall and smart, a natural leader who could be quite entertaining. But because his

Benevolence and Conscientiousness were only average, his strong individual will was apt to lead him astray. If this man were a bad man, it was due to the influences and associations around him.[15]

With the Fowlers' reports published, the stage was set for the anonymous sender to reveal the truth of the originating heads, which happened in the December issue of the journal. The sender said he was happy to report his "unqualified pleasure" in the completion of this "severe test of the truth of phrenological science," which came out of "this fiery furnace like fine gold." The conclusions of the test, Orson added, demanded that "ye doubters" of the science must "either admit that Phrenology can predicate character, or else explain these coincidences satisfactorily on other grounds. No dodging—no backing out, but do one thing or the other."[16]

So who were the owners of the heads? The first cast was taken from the skull of William Teller, a murderer who was executed in Connecticut in 1833. Teller's life was one of crime and imprisonment. Beginning around ten years old, he robbed people and businesses, then was captured and spent time behind bars. Shortly after being released, he stole again, was arrested again, and went right back to jail. This pattern continued for almost two decades, until 1832 when he was sentenced to fifteen years of hard labor. While in prison he led a group of inmates in an uprising and, in the process, murdered a guard. The following year he was executed.

The second cast came from the skull of a Winnebago chief named Big Thunder. Known as a brave warrior, cruel conqueror, and unapologetic marauder, Big Thunder was charismatic and able to lead his Winnebago fighters in great raids on white settlements. In fact, Big Thunder earned his name from the terror he struck in his enemies and the "trembling awe of his countrymen by his gigantic proportions and appalling prowess."[17]

With the truth of these heads on display for everyone to see, the anonymous sender finally revealed his identity: Nelson Sizer. No one knew it yet, but Sizer was on his way to becoming a general in the Fowler empire—the most important nonfamily member in the whole organization. Born in Chester, Massachusetts, in 1812, Sizer grew into a prominent, powerful head. His brow was heavy and serious, his eyes deep and strong. His forehead was high and round, but the sides and back of his head were relatively flat, meaning that his animal propensities were small compared to his intellectual and reasoning faculties.

Sizer first learned about phrenology via Spurzheim's visit to the United States. With no idea that he could make a career in the science, he entered the paper manufacturing business and found himself a young wife. Business and marriage were going well for him in Massachusetts, and he even found time to study phrenology on his own and to give his friends head exams. But when his wife suddenly passed away in 1839, he was at a crossroads. Not only was paper manufacturing uninspiring—or at least more so than usual—but he needed to get out of town and create some distance from the painful memories of his wife's passing. That's when he learned that his neighbors in rural Massachusetts, P. L. Buell and William H. Gibbs, were delivering lectures on phrenology and needed an extra set of hands. Newly alive with purpose, Sizer teamed up with Buell and Gibbs, forming a kind of phrenological trifecta. On some occasions Sizer worked alongside Buell, on other occasions alongside Gibbs, and on still other occasions by himself.

In the fall of 1838 Sizer got his hands on a new periodical that, he believed, would change the course of history. Published in Philadelphia, *The American Phrenological Journal and Miscellany* was "the organ of a great cause," he realized, and it would go much further in spreading the truth of phrenology than lecturing and in-person demonstrations combined. He resolved then and there that the next time he was in Philadelphia, he would visit those responsible for the world-changing periodical and pledge his fealty to their crusade.[18]

When Sizer finally arrived at the Phrenological Atheneum in 1839, he had the pleasure of meeting not just one Fowler sibling but all three. On that particular day, Lorenzo was visiting from New York, and Charlotte was still working at the Athenaeum with Orson. Also present was the first editor of the journal, Nathan Allen. Here, Sizer knew, was the royal court of practical phrenology; he

Nelson Sizer (1812–1897), the most prominent figure in the Fowler empire who was not a member of the Fowler family

pledged to them that he would "aid and strengthen" the science by whatever means necessary.[19]

Over the next decade, Sizer worked as a practical phrenologist, traveling the country, lecturing to countless audiences, examining tens of thousands of heads, and selling subscriptions to *The American Phrenological Journal* (while taking a cut for each subscription he sold). When, in 1846, he mailed Orson a box with two unknown casts in it, he was doing his part to further the truth of the new science. He knew that Orson would nail the test and prove that phrenology was a legitimate, replicable science. Slowly but surely, Sizer proved himself to be a smart, strategic, hardworking lieutenant in the Fowlers' crusade for truth.

WHEN ORSON TOOK stock of the practical, American, democratic direction he had given the journal, he was thrilled with the results. By the end of volume four, during Orson's first year as editor, subscribers had increased from 400 to 900. By the mid-1840s, that number was at 5,000. There were 20,000 subscribers in 1848, and in the 1850s, the number leveled off at 50,000. This was not as many sales as the leading periodicals of the day, such as *Harper's New Monthly Magazine* and *The Atlantic*, but they were still impressive mass-market numbers.[20]

With *The American Phrenological Journal* headed in the right direction under Orson's editorship, and with Lorenzo continuing to move tens of thousands of copies of *The Phrenological Almanac*, the siblings knew it was time to establish a more permanent publishing business. Previously they had dabbled in different arrangements, especially given their proximity to printers in Clinton Hall and the surrounding neighborhood. They first used a variety of contract printers to bring their work to market. Then they formed a partnership called "O.S. & L.N. Fowler," but the name seemed limited and lacked the gravitas they wanted. Plus, Orson and Lorenzo were already stretched too thin with their lectures, travels, writings, and phrenological exams. Running a publishing firm properly would be too much.

The future of their venture finally became clear in 1843 when they partnered with a young man named Samuel Wells. Bright-eyed and kindly looking with a well-defined brow and high sloping forehead, Wells was born in 1820 and raised on a farm along the banks of Lake Ontario. As a young man he dreamed of pursuing a degree in medicine and even set aside a few hundred dollars to attend Yale medical school. But everything changed in 1836 when Wells, still a teenager, traveled to Ithaca with his family and came across a phrenological chart that had been marked the previous year by a woman named Charlotte Fowler. He was taken by the chart immediately but had no inkling of the future significance of the woman who had made it. A few years later, when Wells was living in Portland,

Samuel Wells (1820–1875), husband of Charlotte Fowler and business leader of the Fowler empire

Maine, he heard about an upcoming series of lectures by the Fowler brothers soon to take place in Boston. He made his way to the big city, listened to their lectures in awe, and "his mind became so absorbed with Phrenology that he determined to be a student of the Fowlers." So he packed up his belongings and relocated to New York City, where he began visiting the Phrenological Cabinet daily to learn about the wondrous new science.[21]

Orson and Lorenzo were impressed by the young man's scientific mind, while Charlotte was impressed by the whole package. Six years his elder, she found Wells brilliant, charming, and warm. They married in 1844. Though they never had children of their own, they would largely control the operations of the Fowler empire, especially in the long run, overseeing the business for decades even after Orson and Lorenzo had moved on.

At first, Wells worked as an apprentice at the Phrenological Cabinet, but the siblings soon saw another role for him and Charlotte in stabilizing their publishing operation. To divide the labor of running a burgeoning empire, Charlotte and Samuel agreed to handle the business side of the firm, while Orson and Lorenzo agreed to handle much of the writing, speaking, and promotion. They named their publishing house "Fowlers and Wells," which represented the two Fowlers (Orson and Lorenzo) and the two Wells (Samuel and Charlotte). Once the partnership was formed, the Fowlers took control of the books they had already published and started flooding the market with new phrenological writings, including books titled *Hereditary Descent*; *The Illustrated Self-Instructor in Phrenology and Physiology*; *Intemperance and Tight Lacing*; *Fowler on Memory*; *Physiology, Animal and Mental*; *Practical Phrenology*; *Religion, Natural and Revealed*; *Self-Culture, and Perfection of Character*; and *Synopsis of Phrenology*.

While Samuel headed up the business side of publishing, Charlotte oversaw daily operations. She was responsible for mailing books, pamphlets, journals, and artifacts to people across the country and for interfacing with typesetters, engravers, printers, bookbinders, and others. At any given time, anywhere from six to twelve people worked under her leadership to get Fowlers and Wells publications out the door. Many of these people were relatives, including the seldom mentioned fourth Fowler sibling, Theon, who seemed to work at the firm whenever he needed steady employment. Other family members, including Horace's brood from his second marriage, worked under Charlotte, along with a smattering of in-laws and cousins. There were many nonrelative employees besides; Charlotte regularly insisted that they always had more applicants than spots they could fill.[22]

In 1851, the firm published an account of their reach, detailing just how many publications they pushed out the door. Charlotte and her crew were responsible for mailing upward of 100,000 items every month. Every year, they shipped 45 million pages, which was on top of the many other activities happening under the auspices of Fowlers and Wells—350 public lectures, 10,000 head examinations, 1,200 character descriptions of around ten pages each, and private classes. Year after year after year.[23]

Book sales were also impressive. Orson and Lorenzo's book *Phrenology Proved* sold 20,000 copies in its first six years of availability. Orson's book on marriage, titled simply *Fowler on Matrimony*, sold 20,000 copies in two years, and his book on memory, titled *Fowler on Memory*, sold 15,000 in one year. His pamphlet *Synopsis of Phrenology* sold 150,000 since it was first published (largely because it often accompanied character readings), while his lecture on temperance sold 12,000 copies. Because "immense numbers" of his phrenological charts also sold well, Orson predicted "almost half a million of his various productions are now in the hands of the American public." This was science for the people, and they were just getting started.[24]

IF YOU WERE feeling ill in the 1840s or 1850s, the remedy might be water. This didn't mean just staying hydrated. To purge your body of toxins, you might soak yourself—inside and out—for extended periods of time. The procedure was known as water cure, or hydrotherapy, and it went like this: You'd begin the day by wrapping yourself in a cold wet sheet, designed to stimulate your organs, including your skin, to draw out impurities. After two or three hours in the sheet, which would cause you to sweat uncontrollably, you'd go straight to a frigid plunge bath. After that, you'd drink as much ice-cold water as you could—at least a half dozen tumblers. Then you'd stand under a "douche" (otherwise known as a shower) that poured ice-cold water over you for at least ten minutes. After that, you'd drink several more tumblers, go into another frigid bath, drink some more, and conclude the day by soaking your feet in cold water. Whenever you weren't in a bath or under a shower, you'd place wet bandages all over your body.[25]

Billed as the secret to true health and well-being, water cure began in Germany in the early nineteenth century, spread to England in the 1830s, and landed in America in the 1840s. To many Americans, water cure was not just a matter of health but also a kind of religious practice. The therapy aimed to purify the mind, body, and soul, leading people closer to wellness and to God. As with phrenology, proponents of water cure preached it with evangelical zeal, creating converts who would, in turn, spread the gospel further.[26]

As soon as Orson heard about water cure, he started experimenting with it. Both he and his wife tried it on themselves when they got sick—and to notable effect. But the biggest test came when their infant son ended up on death's door. Sickly from birth, the baby one day developed a fever and a racing heartbeat, which caused the veins in his forehead to turn a steely blue. Combined with his "very pale and ghastly" face, the boy looked more monster than child. Unhappy with what doctors were trying to do for the lad, Orson took matters into his own hands and started draping the boy in cold wet sheets. Although he had to change the sheets several times a day, the illness was slowly pulled from the child's body. "*The wet sheet* HAS SAVED MY BOY," Orson trumpeted in the pages of *The American Phrenological Journal* in 1844. "I fully believe that without it, he would have died." Unfortunately, it wasn't a long-term solution; Orson's son died three years later.[27]

Fowlers and Wells viewed water cure as a natural complement to phrenology, so when the opportunity arose to expand their publishing efforts, they seized it. In 1845, a new periodical appeared in New York City called *The Water-Cure Journal, and Herald of Reforms.* Founded, edited, and published by Joel Shew, a vegetarian and devotee of the health teachings of Sylvester Graham (inspiration for the graham cracker), *The Water-Cure Journal* explained hydrotherapy techniques, publicized books related to the science, and reviewed facilities that offered treatments. Shew was, of course, quick to recommend his own water-cure facility in New York City, which instructed patients to bring with them "four good woolen blankets, some coarse towels, and syringes." Using supplies from home would help Shew keep expenses low and pass the savings on to customers, who would leave the facility with fewer toxins in their body and a little bit of money still in their pocket.[28]

As it turned out, publishing a journal, running a water-cure facility, and administering treatments was much more difficult than

Shew had reckoned. But he had gotten to know the Fowlers reasonably well since opening his facility in New York, and when they saw Shew struggling, they made a pitch: Fowlers and Wells would publish *The Water-Cure Journal*; Shew would continue as editor, but they would exercise their distribution and marketing muscles to push water cure to new heights. Shew excitedly agreed.

In 1848, when Fowlers and Wells took over, *The Water-Cure Journal* had fewer than 900 subscribers. Knowing it could reach many more people, the Fowlers doubled the size of the journal from sixteen to thirty-two pages yet kept the price the same at one dollar per year. They then publicized it extensively, including in *The American Phrenological Journal*. Lorenzo even published a phrenological reading of Joel Shew, deeming his physical constitution "a sturdy oak, bidding defiance to the storm, and strengthening at every blast"; his mental constitution was "one of great elasticity and power."[29]

It worked. In one year subscriptions went from 900 to almost 20,000, making it the second flagship periodical of the firm. Sensing a robust market for hydrotherapy, Fowlers and Wells then launched the Water-Cure Library, an entire line of publications devoted to the new therapy. A key part of the library was *The Water-Cure Almanac*, which was published yearly, as all almanacs were, and was designed to mimic the wildly successful *The Phrenological Almanac*. Fowlers and Wells also put out a slew of related books, including Joel Shew's *The Water-Cure Manual: A Popular Work* (1849); *The Water-Cure, Applied to Every Known Disease* (1850) by J. H. Rausse; and *Experience in Water-Cure* (1852) by Mary Gove Nichols, otherwise known as the "Prophetess of Health." By 1852, the Fowlers and Wells book list had over twenty-five titles related to water cure.[30]

Seeing even more potential for the treatment, the Fowlers established the American Hydropathic Association, making Joel Shew president and Lorenzo Fowler vice president. The organization was in charge of publicizing the science, preparing the journal, and sharing novel treatments with other practitioners. "There is no system so simple, harmless, and universally applicable as the Water-Cure," announced an advertisement for the association. "Its effects are almost miraculous, and it has already been the means of saving the lives of thousands, who were entirely beyond the reach of all other known remedies."[31]

Although water cure quickly became the second most popular science under the Fowlers and Wells name, the firm also saw exciting expansion opportunities with other new sciences. Chief among them was magnetism, also known as mesmerism, developed by a German physician and astronomer named Franz Anton Mesmer. Mesmer's basic claim was that the universe was shot through with a weightless, invisible fluid that acted magnetically on and in all bodies. "Animal magnetism," he analogized, was akin to electricity but more elemental and subtle. You could actually control the flow of this invisible fluid inside of someone else, without even touching that person. If physicians could learn to act through the fluid, they could effectively perform miracles, assessing, diagnosing, and healing the most horrific illnesses without laying hands on a patient.[32]

Today, mesmerism is most often thought of as a cousin to hypnosis, but in the 1830s and 1840s, it was an all-encompassing system of health, wellness, and reality. It took root in the United States when a young, self-styled professor of mesmerism, Charles Poyen, visited America in 1836. Poyen was a showman brimming with gumption and hubris, performing across New England to entertain and enlighten the masses. After he lectured, he would call volunteers to the stage and place them in a mesmeric trance (much as a hypnotist would today). Though the volunteers wouldn't remember a thing about their time in the trance, they would wake up cured of whatever ailment had afflicted them. Poyen purportedly cured everything from stomach pains to cancer, nervousness to liver disease—all without touching people.[33]

The Fowlers were fascinated with mesmerism as a potential tool of phrenology, although they preferred the term *animal magnetism*, believing that a true natural science should not be tied to a person's name. Orson experimented with it as early as 1836 and started covering it extensively when he assumed editorship of *The American*

Phrenological Journal. In fact, he encouraged all phrenologists to approach the new science with an open mind, and not to dismiss it "with a sneer, not treat it like a humbug." Real scientists, he insisted, should not "dismiss *any* matter unexamined which appeals to *experiment*."[34]

Lorenzo was excited about magnetism's potential in medical procedures. In 1842, he put a lady in Boston under a trance, then one of his physician friends cut a large tumor out of her shoulder. Not only did the woman feel no pain, but she did not know the tumor "had been taken from her, until, on being awaked, she was surprised to find it gone." Over the years Lorenzo participated in countless "magnetic surgeries" like this one, serving as magnetizer while a surgeon friend did the cutting. In an age before anesthesia, magnetism was a godsend to those in pain.[35]

Magnetism's success in surgery got Lorenzo thinking about how he might use it in phrenological experiments. Because phrenology maintained that different parts of the brain performed different functions, the younger Fowler brother surmised that he could act through the magnetic fluid to stimulate particular parts of the brain and elicit specific behaviors. To test the theory, he once magnetized a woman and influenced her "Alimentiveness," the phrenological organ responsible for food and drink. Even though she had had a very poor appetite before the trance, she woke up and became a glutton, eating "more every day than before in any one *week*." On another occasion, a young man had his Combativeness and Destructiveness mesmerically influenced. While in a trance, he sprang from his seat on a sofa and, "with the apparent strength of a giant, and with a power both of muscle and will," picked the sofa up and threw it across the room. He then stormed outside and toppled a two-horse sleigh stocked with water as though it had been a handcart.[36]

Once again seeing a market for the new science, Fowlers and Wells launched a series of publications on the intersection of phrenology and magnetism. The series included *Philosophy of Electrical Psychology* (1850) by John Dods; *Fascination; or the Philosophy of Charming* (1847) by John Newman; and *Elements of Animal*

Magnetism (1854) by Charles Morley. Taking a step back and surveying the scientific progress before him, Orson confidently labeled "Phrenology, Physiology, and Magnetism" the "triple stars" in the firmament of human knowledge. The truth of animal magnetism, he concluded near the end of his career after reportedly conducting 10,000 magnetic tests, "is placed beyond a doubt by experiments which all can make."[37]

THE STRANGE, DYNAMIC world of new science in the pre–Civil War decades grew even stranger and more dynamic when the ghosts started showing up. The spirits from beyond the grave even floated their way into the home of Lorenzo and Lydia Fowler, which they shared with Charlotte and Samuel Wells, along with a rotating cast of half-siblings and other relatives. For a while, it seemed as though the whole nation was haunted by a legion of otherworldly forces. A new age of spiritual reality, wherein the veil between the natural and supernatural realms could be torn asunder, was dawning.

It all started in 1848 in the small, rural hamlet of Hydesville, New York. One evening, in a makeshift wooden home, fourteen-year-old Maggie Fox and eleven-year-old Kate Fox discovered that they could communicate with the dead. Specifically, they were able to elicit "knocks" from the afterlife by posing a series of yes-or-no questions. One knock meant yes, and two knocks meant no. At first the Fox sisters simply wowed their parents with this otherworldly ability. But word of their powers quickly spread to neighbors who wanted to hear from beyond the grave. At times, two dozen people crammed into the small Hydesville home to hear the echoes of ghosts.

Soon word of their abilities spread to neighboring towns, where residents, desperate to connect with their dearly departed, clamored for the Fox sisters' help. In 1849, they made their way to Rochester and summoned the spirits before a crowd of 12,000 people in the city's largest performance hall. Then it was on to New York City, where local newspaper editor and friend of the Fowlers Horace

Margaret, Kate, and Leah Fox, the sisters who sparked the Spiritualist movement

Greeley met with the young mediums in a hotel suite to gauge their powers. "We had scarcely taken seats," Greeley wrote in the *New York Tribune*, "before a succession of raps was heard, some on the table, and some on the floor. They differed in violence, and evidently originated from different sources. Some were so strong as to jar the floor or the top of the table." With Greeley's coverage, the Fox sisters quickly became household names, and their powers in high demand. Writers, politicians, businessmen, reformers, ministers, and ordinary Americans paid them a visit to receive wisdom from another world. Over the next decade, more than one million Americans would profess their belief in communication with the dead and identify as Spiritualists.[38]

Though the Spiritualist movement blossomed because of the Fox sisters, the idea of communicating with the dead had been around for millennia. Ghost communication appears in sacred

A typical nineteenth-century séance circle, held to elicit messages from the spirit realm

scriptures from thousands of years ago, including in the Hebrew Bible. Throughout the Middle Ages and well into the Enlightenment, spirit encounters and otherworldly messages were relatively regular occurrences. Even in America, years before the Fox sisters elicited raps from beyond the grave, a man named Andrew Jackson Davis, otherwise known as the Poughkeepsie Seer, claimed the ability to deliver lengthy lectures on behalf of the departed. But it was the young, pretty, seemingly pure Fox sisters who turned the idea of spirit communication into a social, religious, and scientific movement that captured the nation's imagination.[39]

A couple months after the Fox sisters took New York City by storm, a young man named Edward Fowler moved to the Big Apple to pursue a medical degree. A younger half brother of Orson, Lorenzo, and Charlotte, the sixteen-year-old Edward needed a place to stay in the city, and the mansion at 233 East Broadway always welcomed relatives. As Edward settled into New York and started studying medicine, he also began to feel a ghostly touch. By the fall of 1850, he was ready to experiment with channeling the invisible sages who seemed interested in imparting wisdom.

It's unclear exactly when Edward first channeled the dead, but on November 28, 1850, Charlotte started to log his spirit communications. On this frigid Thursday in late November, she convened a séance circle with Edward as well as Samuel and Almira, who were also members of the second Fowler brood. The group gathered around a large wooden table in the spacious entertaining area of the mansion and attuned their minds to the ethereal world around them. After an eerie calm, a ghostly presence descended on the gathering and rapped out a message:

Put out the lights.

A few moments after they extinguished the candles, Almira was "very much effected magnetically," while Edward passed into the "superior condition"—a trance state that enabled the spirit to speak through him. Edward's body rocked and spasmed as the ghost entered him, but then froze in the chair. After a few moments, he stood up and walked into an adjoining room, where he laid down on the sofa, apparently at the spirit's behest.

The other members of the séance circle looked around, unsure of what to do next. After a pensive few minutes, they stood up, walked into the room, and formed a new circle around Edward on the sofa. Again spasming, he spewed forth an otherworldly message: "The past is an atom to what is before us. God is everything. The earth is a part of God. What do we mean by eternity and the universe? Imagination is far below its conception."

This was the message of J. P. Cornell, a banker who had died in 1849. Cornell, who was unknown to Edward prior to this incident, revealed himself to be the young Fowler's spirit guide, an ethereal being who would provide him guidance and insight amid the tumult of the mid-nineteenth century. What's more, Edward had discovered that, like many other mediums in the wake of the Fox sisters, he could channel much more than raps and knocks; he could deliver philosophical statements on behalf of the other world.

The ghosts of the Spiritualist movement, like the ghost of J. P. Cornell, were neither malevolent nor tricksters. They were helpmates who had a distinct, beneficial view of the messy world on earth. But even calling them ghosts is a bit misleading. They were not really dead but spirits who had departed this realm and were now making their way through the ethereal realm. How did Spiritualists know that? Science. Like phrenology, Spiritualism was a movement to detail the laws of nature. What governed communication between this realm and the next was not some kind of supernatural magic but a decidedly scientific process that was subject to observation, experimentation, and replication, just as any science is. Of course, Spiritualists hadn't yet fully pinned down these natural laws—nature was always revealing new secrets. But one day, talking to a spirit would be as mundane as chatting with a friend over a cup of coffee.

A few days after first connecting with Cornell, Edward again passed into the superior condition to relay new messages. With his siblings around him and the spirit working through him, he rose from the séance circle, grabbed a pen and paper, and walked into the adjoining room, which was completely dark. He stayed in there for about ten minutes, then returned to the group and revealed the spirit's thoughts on social reform. The note read, in part:

> The greatest good which you can now do is not to try permanently to perfect society; not to fling your seed to the waste, but to clear the ground, and prepare it ready to receive the seed, and after the ground is cleared, you will best know what kind of seed it is best adapted for. Look back to the various and systematic changes which society has thus far undergone, and you will at once see your true position.

Enabling the spirits to write messages like this was a practice known as automatic writing, which usually produced broad, amorphous, philosophical messages just like Edward received. Seemingly uninterested in specific directions and clear details, the spirits were more like cosmic fortune cookies, complete with squishy generalities that appeared interesting at first but were, on closer inspection, largely forgettable.

That didn't bother Edward, Charlotte, and members of the

séance circle who gathered regularly at the Fowler mansion. Throughout late 1850 and early 1851, the group met once or twice per week and grew to include several friends and neighbors. On December 4, Horace Greeley himself showed up to the séance with none other than the young Kate Fox, who moved in with Greeley and his wife after the other Fox sisters had left New York. At the séance, with Greeley and Kate holding hands around the table with the others, Edward passed into the superior condition, connected with Mr. Cornell, and wrote down a philosophical treatise on the state of the human soul.[40]

With all these philosophical insights swirling around New York, the Fowlers saw in Spiritualism the same thing they saw in phrenology—namely, the promise of improvement. Spiritualism, phrenology, magnetism, water cure—they were allied sciences of progress and social change. Little wonder, then, that the Fowlers brought Spiritualism into their publication list. In fact, even before the Fox sisters started the Spiritualist movement, Orson and Lorenzo had taken a keen interest in the work of Andrew Jackson Davis, the Poughkeepsie Seer. In 1847, Lorenzo gave Davis a phrenological examination and published the results in *The American Phrenological Journal*. Davis had a "highly active" mind, Lorenzo reported, and was "easily influenced or awakened by external objects or internal emotions." His organs of Hope, Combativeness, and Destructiveness were all fully developed, allowing him to push through obstacles that stood in the way of truth. He spoke plainly and candidly about the realities of the spirit world, doing "full justice to his cause." Yet the most remarkable feature of his head was his "intuitive perception of human nature, of motives, and the result of causes."[41]

When interest in Spiritualism took off in the wake of the Fox sisters, the Fowlers doubled down on their connection to Davis. Orson published a series of articles in *The American Phrenological Journal* about the relationship between phrenology and clairvoyance, insisting that the new sciences fit perfectly together: "Is not the soul endowed with a spiritual entity in perfect keeping with this clairvoyant power? Phrenology says YES." In 1851, as Charlotte was conducting séances with Edward, Fowlers and Wells published a new book by Andrew Jackson Davis, *The Philosophy of Spiritual Intercourse*. It contained chapters titled "God's Universal Providence," "The Miracles of This Age," "The Guardianship of Spirits," "The Discernment of Spirits," and "A Voice from the Spirit-Land." Orson reviewed the book and concluded that "no person can cease from their perusal without having an increase of Faith, Hope, Charity, and positive Knowledge. We commend the volume to all who desire information on these vastly important, but misunderstood, subjects."[42]

More books on Spiritualism, clairvoyance, and insights from beyond the grave followed Davis's book. Fowlers and Wells published *Immortality Triumphant* (1852) by John Dods; *Supernal Theology and Life in the Spheres* (1852) by Owen Warren; and *The Macrocosm and Microcosm; or, the Universe Without and the Universe Within* (1852) by William Fishbough, among other titles. Orson never considered Spiritualism on par with phrenology, physiology, and magnetism in terms of importance for human knowledge, but it was significant enough to play a key role in the output of the publishing giant that Fowlers and Wells was becoming.

FOR HIS PART, Edward continued to channel spirits and to pronounce messages of promise and progress. In August 1851, he, Charlotte, and other Fowler siblings, along with a half dozen compatriots in the Spiritualist movement, formed a group known as the New York Circle, which dedicated itself to "making careful observations concerning modern Spiritual phenomena." The phenomena the group witnessed, all done through Edward's mediumship, included levitating tables and chairs, sofas sputtering across floors, colorful lights dancing through the dark, objects whipping across the room, and an endless stream of messages.[43]

Edward even started writing messages in foreign languages that he did not know. The first was in Spanish, the second in Hebrew, the third in Sanskrit. Finally, on November 24, a spirit showed up to explain what these foreign messages meant.

"I am happy to announce to you," the spirit of Benjamin Franklin told the group, "that the project which has engaged our attention for some years has at last been in part accomplished." According to Franklin, a new Spiritualist era was about to begin, and Edward's foreign-language messages were a beacon light to the world.

A couple weeks later, Edward entered the superior condition again and received a new command.

"Edward," the spirits said, "put that paper on your table, and we will write a sentiment and subscribe our names; then you may sign it too."

Edward did as he was told, placing pen and paper on the séance table before heading off to bed. When he and the others awoke in the morning, a sagacious message had appeared on the paper: "Peace, but not without freedom." It was signed by the founding fathers—Franklin, Jefferson, Adams, Hancock, Washington, and many others. When members of the New York Circle obtained a book that showed what the original signatures of the founders looked like, they excitedly reported that the signatures matched those on Edward's piece of paper. It was important news. The spirits of the founding fathers could see the path the United States was on in the 1850s, a decade that would end in bloody civil war. Peace was the ultimate goal, the founders announced, but to get there the nation first needed freedom for everyone within its borders. Apparently, the founders had become abolitionists in the afterlife.

The New York Circle continued until at least 1854. But once Edward earned his medical degree in 1855 and faced new career prospects, the spirits came around less often, which was probably for the best. Doubts began to swirl around Edward's abilities. Both Samuel Wells and Nelson Sizer were skeptical, although they had no interest in debunking him. At the same time, newspapers and periodicals began lambasting Edward and other so-called Spiritualists for perpetrating obvious frauds. Then, at one point, at least according to a relative on the Wells side of the family, Edward admitted to faking the whole spirit-medium thing. A young man caught up in a lark, he decided to focus on his medical career and to leave the spirits behind.[44]

Yet even as interest in Spiritualism waned among the Fowlers, excitement about phrenology was reaching a fever pitch, bringing a cavalcade of celebrities around the family—and making them celebrities in their own right.

CHAPTER SIX

CELEBRITY SCIENCE

NEW YORK'S CHIEF OF POLICE ARRIVED AT THE VENUE three hours early, leading a column of fifty men to the front of Castle Garden. The cops formed a tight line from State Street, where the carriages lined up, to the venue itself, which was accessible only via a walkway that connected Battery Park to the little island on which Castle Garden stood. When the carriages pulled up, elegantly dressed New Yorkers hopped off and strode through the phalanx of police, completely aflutter of what they were about to experience.

Despite the throngs of gawkers who could not afford a ticket but still wanted to catch a glimpse of this historic moment, no real violence broke out. The biggest disruption came from the hooligans in the harbor surrounding Castle Garden. Packed together on rickety boats and drunk off alcohol and oysters, they strained their ears to hear whatever they could. When that didn't work, they hooted and hollered, playing fifes and drums as loud as they could to draw attention away from the performance. At one point, some of the hooligans even tried to make landfall and storm Castle Garden, but police were able to push most of them back. The few who got on land were quickly arrested.

Inside the venue, the elegantly dressed New Yorkers could hear some of the hooting and hollering outside, but it was the spectacle before them that truly captured their attention. Brightly colored lamps marked different seating sections and created a pulsating visual display across the auditorium. The interior columns and balcony were adorned with resplendent works of art trucked in just for this occasion. Hanging from the ceiling were massive chandeliers that refracted colorful light from the lamps and seemed to dance and sparkle in midair. Then there was the stage itself, expertly crafted atop a platform built over the orchestra pit and reinforced with a sounding board instead of a curtain to improve acoustics. A sign made of freshly picked flowers hung from the balcony and announced: "Welcome, Sweet Warbler."

Finally, at eight in the evening on September 11, 1850, the program began. There were two opening acts. Julius Benedict conducted a piece from his opera *The Crusaders*, then Giovanni Belletti, the renowned baritone, sang songs from *Maometto Secondo*. But these performances were simply prelude to the main attraction, which brought forth a moment of "breathless expectation." Adorned in a simple yet stylish white dress, her hair braided, her makeup delicate, the Swedish Nightingale stepped onto the stage carrying a hefty bouquet. Then, pandemonium. Thunderous applause lasted for several minutes as the crowd waved hats, hands, and handkerchiefs and threw flowers at her feet. The "divine songstress" looked out at the thousands before her with "that perfect bearing, that air of all dignity and sweetness, blending a childlike simplicity and half-trembling womanly modesty with the beautiful confidence of Genius and serene wisdom of Art."

Jenny Lind, the Swedish Nightingale, had experienced scenes like this countless times before, but only in Europe. This was her first time in the United States, and it felt incredible. After the crowd quieted, she began to sing. For the next ninety minutes she enchanted New Yorkers just as she had enraptured the aristocracy of the Old World. Her voice—"so pure, so sweet, so fine, so whole and all-prevailing"—enveloped Castle Garden. One paper called her "the greatest prodigy in song that ever appeared upon the theatre of the world." Hers was a voice that "cannot be described" but "must be heard" to be understood.[1]

When Lind brought the performance to a close, the crowd

clamored for more. But surprisingly it wasn't for more songs—it was for another person. "Where's Barnum?" they roared.

P. T. Barnum had engineered the evening to perfection. Everyone knew that he was responsible for bringing Jenny Lind to America, a point that he made sure the papers included when covering the concert. Although he had long used the press to drum up attention for his schemes, this occasion was different. Jenny Lind was his first high-brow attraction, his attempt to change his reputation from a perpetrator of humbuggery to a purveyor of taste and class. And the nation loved him for it.

When Barnum heard the audience clamoring for him, he smiled brightly, marched onto the stage, and took his place next to Lind. "I have but one favor to ask of you," Barnum announced, "and that is, that in the presence of that angel, I may be allowed to sink where I really belong—into utter insignificance." It was a great way to call attention to himself by appearing to deflect attention away from himself. But Barnum couldn't stop there; he had an important announcement to make. Just that morning, Lind had told him that she would not accept her portion of the profits from the performance, estimated to be around $10,000 (almost $400,000 in today's dollars). She would instead "devote every farthing of it tomorrow morning for charitable purposes." Pandemonium once again. Lind's performance and selfless generosity ushered in a new "epoch in the musical annals of America," the *Brooklyn Daily Eagle* concluded.[2]

Thousands of people had packed inside Castle Garden to hear Jenny Lind sing. One was Lorenzo Fowler. As was his wont, he surveyed the heads around him as he waited for the show to begin. What were the characters of the people fortunate enough to buy tickets to this epoch-making event? He noticed one man in particular seated just down the row. The man's head, Lorenzo later recalled, "was predominating in the upper portion," and he seemed to vibrate with nervous excitement before the concert started. Once the Swedish Nightingale began her siren song, the man's nervous energy dissipated, and he let the music wash over him. Ninety minutes later, Lorenzo looked down the row and saw the man grasping his head. "Oh, my God," the man exclaimed, "what an angel!" It was a perfect choice of words. Lind's voice had "stimulated his whole moral frame," the phrenologist reflected. Indeed, her singing "awakens sentiment, and makes one feel that he has a soul, because she has Veneration and Benevolence; and when she sings she becomes transported almost beyond this world." Perhaps she was an angel.[3]

A year after listening to Lind at Castle Garden, Lorenzo published a detailed reading of her in *The American Phrenological Journal*. In one of the longest profiles printed up to that point, he described Lind's moral and social organs as surprisingly well developed, which meant that she sang not for herself, or for wealth, but for the improvement of her audiences—"radiating the light and warmth of her joyous nature upon others." Even more significant were the organs that structured her sturdy brow line. With "a great prominence of the outer angle of the eye-brows," which creates a "bold ridge" above the eyes that "makes the forehead in that region appear deformed," Lind's Order, Number, Time, and Tune resulted in an unparalleled voice.

Yet equally as important as her voice was her Veneration and Spirituality, which attuned her talents to the musical connection between the heavens and the earth. With "a decidedly religious cast to her mind," her "whole soul and body" come together "to make up her musical talent." Jenny Lind was the greatest singer of nineteenth century, and perhaps of all time, because her entire constitution—from her temperament to her organs to her mental functions—meshed perfectly together.[4]

To Lorenzo, Jenny Lind was unqualified proof of phrenology, and not just because of her mental faculties. True, she was naturally endowed with musical abilities, but she also worked hard at her craft, overcoming many setbacks on her road to superstardom—a point Lorenzo made clear in the biographical sketch that accompanied her phrenological profile. Through focused mental exercises, she improved her nature to the point where Order, Number, Time, and Tune pushed out on her brow. At the same time, she advanced her relationship with God and enhanced her moral organs so she could take her listeners on a spiritual, otherworldly journey.

The Fowlers relished the opportunity to promote phrenology through the success of celebrities such as Lind—famous figures who got to where they were because of how they exercised their mental faculties, grew their organs, and improved their natural endowments. Indeed, countless public figures appeared in Fowler lectures, demonstrations, and publications as proof of the science's power. And the American people ate it up, devouring story after story about famous individuals who visited the Phrenological Cabinet, who embraced phrenology, and whose achievements proved the exact ideas the Fowlers were peddling. It was a brilliant strategy for building their brand and supporting the science, as it placed phrenology at the forefront of the rapidly emerging celebrity culture of the time—the first true celebrity culture in American history.[5]

WHEN LORENZO BOUGHT a ticket to Castle Garden to hear the heavenly voice of Jenny Lind, he wanted to experience the music and to support his friend. Barnum and the Fowlers had known each other for at least a decade by that point, and there was a strong bond between them. Although some denounced Barnum for his hoaxes and hoodwinks, all in the name of profit, the Fowlers saw him as an honest broker of the strange, unfamiliar, mysterious side of life. They were interested in much the same thing. Indeed, Barnum and the Fowlers aimed ultimately at engaging the masses. Barnum did it with oddities; the Fowlers did it with science. Both did it in the name of democracy. And they did it next door to each other.

If you were standing near the corner of Broadway and Ann Street facing north in the fall of 1842, you would have seen the Fowlers' Phrenological Cabinet on your right—its rows of skulls staring out at you from the first floor of Clinton Hall. To your left you would have seen Barnum's American Museum, adorned at the time with an eighteen-foot-tall banner announcing perhaps the most shocking discovery in human history: Mermaids were real. The banner depicted beautiful, bare-breasted maidens with long flowing hair and fish tails. Their lusty countenance beckoned visitors inside. Yet once curious spectators got through the doors, they quickly discovered that Barnum's mermaid, which he dubbed the Feejee Mermaid, did not match the busty fish ladies on the banner. A small, shriveled, Frankenstein-like monstrosity, the Feejee Mermaid had been handcrafted from parts of a monkey, ape, and salmon. Barnum had bought it from a shady exhibitor in Boston, knowing he could make a killing by getting people to pay twenty-five cents to see it. And he did. Even if museum-goers departed his business miffed at the hoax, they had already paid him, and he was hard at work concocting another oddity that would bring them back.[6]

If people were aggrieved about losing their quarter to see some ridiculous monkey-fish horror, they could always head across Ann Street to the Phrenological Cabinet. Admission there was free. Surrounded by skulls, they could touch the truth of humanity all day long. And that was one place the Fowlers had an edge over Barnum—tactile reality. At the Cabinet, people could feel the smooth peaks and valleys of a human skull and discover what those surfaces meant. At Barnum's American Museum, they could look at oddities with a mixture of curiosity and disgust, but they couldn't touch what they saw. Of course, that didn't stop people from showing up at Barnum's museum in droves; it was the top tourist attraction in New York. But the Phrenological Cabinet was right behind, emerging as the second-leading tourist attraction in New York during the 1840s and 1850s.[7]

Little wonder that Barnum and the Fowlers struck up a friendship and supported each other. Barnum visited the Phrenological Cabinet for a head reading on several occasions. And in one of his biographies, he thanked the Fowlers profusely for supporting his work as a temperance advocate. (Both Barnum and the Fowlers were proud teetotalers.) Calling Orson and Lorenzo "my worthy friends," he praised their publishing firm for being "pre-eminent in the Phrenological line," and for doing much "to enlighten the public on Temperance, Physiology, and other important matters. Few men have published a larger number of useful books."[8]

For their part, the Fowlers regularly publicized happenings at

Sleighing in New York, *an 1855 painting by Thomas Benecke depicting a carnivalesque winter scene in front of Barnum's American Museum*

the American Museum; increased traffic to the area benefited the Cabinet too. And his exhibits served as proof of their science. Because of their friendship with Barnum, the Fowlers were able to talk with and examine the human oddities who appeared regularly at the American Museum. Exploited for profit, dehumanized and ostracized, the people Barnum exhibited never received the dignity they deserved, but they were central to the celebrity culture that defined Broadway. The most significant human oddity of the day, given the money he commanded, was Charles Stratton, better known as General Tom Thumb. Standing three feet, four inches tall and weighing approximately fifteen pounds, Stratton became almost as famous as Barnum himself. Getting to know him well over the years, the Fowlers understood the challenge that Stratton posed to phrenology. Given his tiny head, and given the fact that brain size correlated to mental abilities in phrenology, how could the new science account for Stratton's humor, wit, charm, and ongoing success?

The answer, Lorenzo explained, was balance: "The size of his head is in harmony with that of his body." Phrenology was not about absolute sizes in all cases; it was rather about relative sizes based on individual people. To prove as much, Lorenzo detailed the sizes of Stratton's organs and concluded that he was "well put together and of good material," with a strong social nature and prominent intellectual organs. "He is very observing, has a good memory of persons, places, stories, and events, can converse fluently, and has quite a taste for music; but he is truly indebted to his Imitation for his successful performances." Phrenology thus explained how a dwarf, born to poor parents in Connecticut, could rise to the top of global entertainment and charm the world.[9]

Other performers at Barnum's American Museum who served as proof of phrenology included Josephine Clofullia, the most famous "bearded lady" in the world. The Fowlers found her "ambitious, fond of display, regardful of personal appearance, and anxious to please." Even though she sported a thick beard, she was a "dignified and self-possessed" woman of refined character. Another was Sylvia Hardy, the "Giantess of Maine," who was publicized as being eight feet tall but was at least six inches shorter than that. Calling her phrenological development "remarkable," the Fowlers were gobsmacked by her "unevenly developed brain," where some of her organs were "expectedly small" and with "limited influence in character," while others were "immensely large and controlling." On the whole, however, "She is independent, quite persevering, and most decidedly kind and generous."[10]

Perhaps the most important performer at the American Museum the Fowlers were able to examine was Barnum himself. If phrenology had anything to say about improvement and success, it should be able to do so via the cranium of the Greatest Showman on Earth. In 1852, the Fowlers published a lengthy account of Barnum's phrenological development, along with a detailed biographical sketch. Praising his "active temperament, with sufficient vitality to enable him to perform much mental or physical labor," the Fowlers printed a chart of the sizes of all his organs. Particularly striking was how many of his organs were "large" (size 6)—Amativeness,

Philoprogenitiveness, Adhesiveness, Alimentiveness, Firmness, Hope, Veneration, Benevolence, Sublimity, Mirthfulness, Individuality, Agreeableness, Human Nature, and several others. This many large organs meant that Barnum had a brain perfectly fitted to the wonder shows he put on across the United States and Europe. Indeed, it was his brain that made the wonder shows possible.

But the Fowlers went further than simply publishing Barnum's phrenological character. They told his life story—how, through hard work and concerted activity, he rose from obscurity to international fame. Born in Danbury, Connecticut, to a middle-class family, Barnum tried his hand at various businesses. He was a clerk for a time, a dry-goods merchant for a time, a newspaperman for a time, and a politician for a time. He failed at all of them. But his "thorough independence, characteristic of his whole life," kept him—and his mind—going, until "his untiring industry, his knowledge of human nature, his aptitude for business, realized him a considerable fortune."

After telling Barnum's story as a rags-to-riches tale, even though such a tale wasn't strictly accurate—Barnum came from a fairly distinguished family in Connecticut—the Fowlers dedicated several pages of their journal to defending their friend from accusations of charlatanry. His hoaxes, they explained, were not the result of duplicity but grew out of honest mistakes. "The *truth* in regard to these and kindred subjects," the Fowlers told their readers, "has never appeared in print, and consequently Barnum has been viewed in a *false* light—a light, perhaps, of his own creating. Nevertheless, as we wish to show him as he is, we must be pardoned for publishing the *facts*, as we know them, briefly, and we repeat, for *the first time*." It was Barnum, they wrote, who had been duped into buying the Feejee Mermaid from an unscrupulous sailor; it was Barnum who had been tricked into believing that Joice Heth was 161 years old and the boyhood nursemaid of George Washington; and it was Barnum who had been deceived into believing the odd-looking "Woolly Horse" he purchased was captured by Colonel John Frémont in the wilds of California.[11]

Why defend Barnum at such lengths? Part of the answer was friendship and common cause. But there was something else. The Fowlers were expanding phrenology through tales of celebrities and successful public figures, whose greatness grew out of hard work and mental improvement—the core promises of phrenology. If Barnum's success was due more to humbuggery than to honest toil in service of greater understanding, then phrenology's story of improvement was suspect, at least when it came to their friend and fellow entrepreneur. Defending Barnum was a way of defending phrenology and the promise of democracy—the promise that everyone could improve their lot in life simply by following the principles of the new science. America was a land where anyone could become a celebrity, including someone like P. T. Barnum.

BARNUM WAS THE most successful purveyor of human oddities in the nineteenth century, but he was just one of many in an era of carnivalesque performance that helped the Fowlers amass proof of their science. In fact, years before Barnum hit the national stage, a traveling oddity show traversed the nation, raking in big profits and establishing the template for future exhibitors. In sprawling cities and small hamlets, people turned out in droves to witness a spectacle and to contemplate some big questions: Was it real? Was it a hoax? Why would God create two brothers who were inseparable, literally, due to a twelve-inch band of flesh on their abdomens?

Chang and Eng, the original Siamese twins, came to the United States after their Chinese mother, who was living in Siam (now Thailand), "leased" them to a ship captain who was passing through the area. The captain, Abel Coffin, saw them playing on the banks of a river and, after approaching them, knew they could bring in the big bucks. And he was right. Inquisitive gawkers across America eagerly paid twenty-five cents for a chance to see the twins and converse with them (after they learned English).[12]

In 1836, after years of performing, the twins announced their return to New York, one of the first cities in which they appeared after moving to the United States. Orson heard the news of their return and, along with his assistant, Samuel Kirkham, dashed to the stately Washington Hotel to convince Chang and Eng they needed a reading. Years earlier the twins had heard of the new science and had even received an examination from an amateur phrenologist, whose account did not impress them. But Orson was different, and Chang and Eng could tell as much, so they let him feel their heads. With the Siamese twins seated on a sofa, Orson stepped behind them and moved spryly across the bumps, spans, and masses of their heads—one hand working on Chang, the other on Eng. As he went, he called out to Kirkham what he felt, and Kirkham dutifully took notes. Right away Orson knew he was touching "a striking example of the truth of phrenological science." In the size of their heads and the details of the characters, the twins were nearly identical, both biologically and phrenologically. "Some small difference, indeed, in the development of some *few* of the organs does exist," Orson acknowledged, but the differences were minuscule compared to the remarkable similarities. "Among all the heads ever examined by the Editor," Orson wrote years later for *The American Phrenological Journal*, "such an agreement in size, shape, and temperament, or anything *approaching* to it, in any two, they never before witnessed or heard of." When this account appeared in the journal, it was accompanied by an attractive engraving of the twins, dressed immaculately as antebellum gentlemen and linked by the fleshy patch of abdomen that amazed the world.[13]

Shortly after Orson's examination, the twins retreated from the national spotlight and settled in North Carolina, where they bought a farm, owned slaves, married local sisters Sarah and Adelaide Yates, and had twenty-one children between them. But public exhibitions, and the money that came with them, were too alluring to leave behind forever. In 1854, after more than fifteen years in private pastures, they agreed to appear for several weeks at the Broadway Museum and Menagerie, an upstart museum meant to rival Barnum's. The museum was opened by Herr Driesbach, a German American performer who had made a name for himself as the best lion tamer in the world. Traveling to far-off lands, Driesbach amassed a zoo of exotic animals—rhinos, hyenas, monkeys, apes, lions, tigers, bears, and more—that he decided to show off to the people of New York. To compete with Barnum, however, he needed more than just animals to attract tourists, so he booked some human oddities, including the Siamese twins and a man named Sanders Nellis, who was born without arms but could shoot a bow and arrow with his feet and play a variety of musical instruments.

Chang and Eng appeared at the Broadway Museum alongside two of their children, whose presence raised unspoken questions of sexual encounters at a time when intercourse was a topic too unseemly for public discussion. As soon as the Fowlers heard that the twins were right up the street, they made quick plans for another examination, hoping to discover what, if anything, had changed in their mental development. Laying hands on the world-famous twins, Orson discovered that the head of one twin was "a quarter of an inch larger than that of the other, and his individual organs correspondingly fuller; but otherwise, we could not detect the slightest difference." This was not terribly surprising, given what Orson had found in 1836, but it was nonetheless significant: "This complete correspondence, considered in connection with their perfect similarity of character, corroborates, as far as one marked fact can do, the truth of our science."

The Fowlers went on to note that their Combativeness was well developed, but they lacked any semblance of Destructiveness or Secretiveness. They had incredibly large organs of Benevolence and Veneration, especially considering their national heritage, where people were often underdeveloped in these areas. As dedicated family men, their Adhesiveness, Philoprogenitiveness, and Inhabitiveness were properly large. It was a point hammered home by another engraving, which featured Chang and Eng, now older and wiser, wrapping their arms around two of their children, the kids leaning comfortably over their fathers' knees.[14]

BECAUSE OF THE pervasiveness of celebrity culture at the time, the Fowlers saw a huge opening for their cause. Practically every periodical, book, or tract they published included vignettes of the rich and famous from every sector of American life. Actors, artists, musicians, novelists, poets, religious leaders, politicians, scientists, doctors, inventors, business tycoons, and more—the Fowlers covered them all to prove that phrenology could account for greatness in whatever way it appeared.

Some celebrities they featured were close friends. Newspaper editor, ghost hunter, and political visionary who helped found the Republican Party, Horace Greeley received several phrenological exams over the years, including one in 1847 when the Fowlers noted his head "combines great strength of mind with a high order of intellectual capacity and mental worth." Another friend was Horace Mann, a prominent politician, reformer, and creator of America's public school system. He visited the Phrenological Cabinet in 1853 and sat for a reading that revealed remarkable Firmness, which had been "unusually stimulated to activity by circumstances, as if his course of life had been a pioneering one."[15]

Other important figures in the Fowler carnival were not family friends, but they were useful for promoting phrenology. Henry Clay, a congressional representative from Kentucky, Speaker of the House of Representatives, and arch nemesis of Andrew Jackson, did not run to the Fowlers for an exam, but he did allow Orson to feel his head, which was remarkable for its unity: "Every organ helps its fellow and contributes to the general result." Then there was Alexander Campbell, renowned minister, theologian, and leader of the restoration movement, which led to the creation of the Church of Christ and Disciples of Christ denominations. In 1847, Campbell visited Lorenzo for a reading but withheld his identity as a test of the new science. He sat in the exam chair gobsmacked as Lorenzo described "with such remarkable exactness what I knew of myself." After leaving the exam, he wrote Lorenzo a letter professing his belief in phrenology and declaring that it would, in short order, take its rightful place as the preeminent science of human nature. There was also inventor Samuel Morse, whose telegraph revolutionized communication. In 1848, he allowed Lorenzo not only to examine him but also to take a bust of his head for posterity. Lorenzo found that Morse's perceptive faculties, Constructiveness, and Ideality were the controlling features of his mind, giving him "that scientific, experimenting, and practical cast of mind which exactly fit him to invent the telegraph."[16]

There was another category of celebrity who were caught up in the Fowlers' phrenological crusade, and that was people who were not yet celebrities but would become so later on. Future leaders who received an exam when they were young included President James Garfield, Civil War hero and President Ulysses S. Grant, and opera singer Clara Louise Kellogg, among many others.

Then there was the future celebrity drawn into the march of phrenology via Lorenzo's lecture in Springfield, Massachusetts, in 1849. Sitting in the audience, enraptured with what he saw and heard, a twenty-eight-year-old John Brown Jr. was awestruck as Lorenzo asked for volunteers from the audience for a public examination, then proceeded to describe their characters perfectly. At the close of the lecture, as was common at the time, a committee of audience members was assembled to pass resolutions thanking the speaker for his contributions to their collective knowledge. The committee in Springfield thanked Lorenzo for restoring confidence in phrenology, for vindicating it from charges of materialism and infidelity, for showing that phrenology is "essential to a more complete and successful study of man," and for instructing them on a science that is "eminently calculated to promote our health, happiness, and usefulness as individuals, and the welfare of society generally."[17]

John Brown Jr. signed his name excitedly to the statement. Over the next decade, Americans would canonize or demonize the Brown family—there seemed to be no middle ground—especially its patriarch, John Brown Sr. In 1856, the Brown family patriarch used a cache of broadswords to hack to pieces several proslavery settlers in Kansas. Then in 1859, he led a raid on the federal arsenal at Harpers

Ferry, Virginia, hoping to set off a revolution that would bring an end to slavery. It didn't work, and on December 2, 1859, he was hanged for treason, murder, and insurrection. Brown's actions were, in many regards, the flame that lit the fuse of civil war, changing America forever.

John Brown Jr. worked alongside his eccentric father to abolish slavery, although he did not participate directly in the Kansas massacre or the raid on Harpers Ferry. He worked logistics to support those missions, but he did not take up arms. And in 1849, he had no idea of the role his family would play in fomenting war; he was simply looking for a career.

John Brown Jr.'s initial interest in the science likely began when his father received a phrenological exam in 1847 from Orson. The elder Brown swung by Clinton Hall while visiting New York and paid for a written description, which John Brown Jr. kept safe for decades after his father swung at the end of a rope. Orson found Brown's character remarkable: You are "very active, both physically and mentally, . . . are more known for your practical off-hand talent than for depth and profundity of comprehension." Orson then continued:

> You have a pretty good opinion of yourself—would rather lead than be led—have great sense of honor, and would scorn to do anything mean or disgraceful . . . You are too blunt and free-spoken—you often find that your motives are not understood . . . In your character and actions you are more original than imitative, and have more taste for the useful than the beautiful and ornamental.

It was a fitting, prescient description of a leader who spoke boldly, almost crazily, of ending slavery, and whose motives would be greatly misunderstood across the country.[18]

This reading, in conjunction with Lorenzo's 1849 Springfield lecture, convinced John Brown Jr. to make phrenology his life's work. Several months after hearing Lorenzo speak, Brown wrote a letter to

Samuel Wells about a recent suicide of "an Englishman, named Gaylord" in the town of Springfield. Morose and melancholic for quite some time, Gaylord had often contemplated suicide, and finally, at five o'clock one morning, he followed through by putting a gun loaded with buckshot in his mouth and, using a long metal rod, depressing the trigger. The buckshot sprayed out the back of his head, painting the room with blood, brains, and skull fragments. It was a "horrid affair," Brown wrote to Wells. Yet it presented an important scientific opportunity for phrenology, and Brown meant to seize it. Waiting around the scene of the suicide until the coroner had finished his job but before local officials hauled the body away, Brown "proceeded to an examination of his head . . . amid blood and mutilation."

Giving a phrenological examination to a head exploded with buckshot was no easy task, but the son of the infamous "Osawatomie Brown" was willing to try. As he approached the corpse, he noticed that the back of the head was "so mutilated that I could obtain but a very imperfect idea of it." But he considered himself a man of science, so he knelt next to the body and moved his fingers across the crimson mess that matted Gaylord's hair. He discovered that Gaylord was extremely deficient in "the reflective organs," with moderate Self-Esteem, Hope, and Benevolence; his Firmness, Veneration, and Conscientiousness were moderate. Only his Acquisitiveness—or desire for money and property—was large. This, Brown concluded, was a constitutional recipe for suicide. "I would take a cast of his head if I knew how to do it," Brown told Wells. But it was not to be. Local officials arrived to claim the corpse, and Brown left the scene smeared with blood and brains.[19]

Out of this gory phrenological exam blossomed a long and rewarding relationship with the Fowlers. After corresponding regularly with Orson, Lorenzo, and Samuel Wells, John Brown Jr. moved to New York in 1850 to begin working at the Phrenological Cabinet. His soon-to-be-famous father even popped in one day to check on his son, only to find that Junior was away from the office. Brown's wife, meanwhile, agreed to supervise the housekeeping for Lorenzo's family. Later that year Brown worked as Nelson Sizer's assistant during a lecture tour in Ohio, spending three months together speaking, examining heads, and impressing audiences. At the end of the tour, John Brown Jr. stayed in Ohio to continue speaking on and promoting phrenology. "The public," the Fowlers wrote of Brown, "may rely upon his integrity in all respects, and confide his skill as a phrenologist."[20]

At some point in 1851 John Brown Jr. went back to New York to work again at the Phrenological Cabinet, where he was classified as a "student and assistant," according to *The Phrenological Almanac* of that year. But his stay lasted only a couple years, after which he went back to Ohio to resume lecturing. "We commend Mr. Brown to the inquiring minds," wrote a committee in Hartford, Ohio, after Brown lectured there in 1854. He is "a gentleman who, by the attractive manner in which he imparts needed and important information, . . . will amply compensate all who may" listen to him. Later that year John Brown Jr. was appointed justice of the peace in Lindenville, Ohio, where he continued to lecture on phrenology and physiology. Then in 1855 he moved with his brothers to Kansas to join their father in battling the so-called border ruffians who wanted to make the territory a slave state. Even though he did not participate in the Kansas massacre, John Brown Jr. was apprehended and tortured by proslavery settlers until John Brown Sr. agreed to trade prisoners of his own for the release of his son.[21]

For their part, the Fowlers cheered on the abolitionist work of their associate. Even though they didn't agree always with the Brown family's methods of agitation, they applauded the fight to end slavery; destroying the evil system was an extension of phrenology's promise of improvement for all. At one point, the Fowlers even sponsored men who relocated to Kansas as part of the free-soil, antislavery movement. They were forever connected to the Brown family and to the flame that lit the fuse of war.[22]

DESPITE HIS CLOSE work with the Fowlers, John Brown Jr. was not, in fact, the most famous American to use phrenology to climb the ladder of greatness. That distinction belongs to the man who entered the Cabinet for a reading on July 16, 1849. Feeling the man's head, deciphering his character and prospects, Lorenzo was amazed. As he told the man,

> You were blessed by nature with a good constitution and power to live to a good old age. You were undoubtedly descended from a long-lived family. You were not (like many) prematurely developed—did not get ripe like a hot house-plant but you can last long and grow better as you grow older if you are careful to obey the laws of health, of life and of mental and physical development. You have a large sized brain giving you much mentality as a whole . . . You are independent, not wishing to be a slave yourself or to enslave others. You have your own opinions and think for yourself. You wish to work on your own hook, and are inclined to take the lead . . . You are a great reader and have a good memory of facts and events much better than their *time*. You can compare, illustrate, discriminate, and criticize with much ability. You have a good command of language especially if excited.

Both Lorenzo and the man came way from this examination convinced that this was the head of a true poet, perhaps the ideal poet, whose "good command of language" could advance the literary world in incredible, unforeseeable ways.[23]

At the time, the twenty-nine-year-old sitter was working as a printer and newspaper columnist, but he would indeed become what the phrenological reading foretold. The ideal poet, the poet of American democracy, he was Walt Whitman, and he would reprint this phrenological reading several times throughout his life as an acknowledgment of the prophetic-yet-scientific power of practical phrenology.

By the time Whitman ventured to the Phrenological Cabinet for a reading in 1849, he was well acquainted with the science. In

March 1846, when he was editor of the *Brooklyn Daily Eagle*, he attended a lecture series by Orson Fowler that gave him a lot to ponder, particularly about the speaker. Describing the lecture as "a most curious and unique jumbling together of phrenology, physiology, physiognomy, religious dialectics," and more, Whitman could not decipher an eloquent through line to the presentation. The head examinations that concluded the lecture were "immeasurably funny," he wrote. "After picturing out the character of the subject," Whitman said of Orson, "he had a curious way of asking him or her whether he has not been correct, as if any person can know his own character or peculiarities."[24]

Though Whitman was unconvinced about phrenology after Orson's first lecture, he continued to attend the series, and his opinion slowly changed. "These lectures, certainly, are the most curious that we have ever attended. Not only does the lecturer profess to give the most accurate delineation of individual character, but he gives the most minute instructions as to the faculties which require development or restraint." Still marveling at Orson's bold lecturing style, which included "the most perfect contempt for the understanding of his audience," Whitman left the lecture hall intrigued by the science but still unimpressed with Orson: "We do not mean to assert that there is no truth whatever in phrenology, but we do say that its claims to confidence, as set forth by Mr. Fowler, are preposterous to the last degree."[25]

With the seeds of phrenology planted firmly in his mind, Whitman published several subsequent notices about phrenological works, including a new edition of Spurzheim's writings and a book called *The Use of the Body in Relation to the Mind* by George Moore. Then in March 1847, Whitman made a rather startling admission: "That the study of physiology is good, admits not denial," and "there can be no harm, but probably much good, in pursuing the study of phrenology." After praising two new books by Fowlers and Wells, Whitman called Orson, Lorenzo, and Samuel "the most persevering workers in phrenology in this country." They must "certainly be reckoned with," he concluded.[26]

Reckon with Fowlers and Wells was exactly what Whitman did, including by making several trips to the Phrenological Cabinet to learn about the new science. Calling it "one of the choice places of New York," he "went there often" and amazed over "the busts, examples, curios and books." It was around this time that Lorenzo gave Whitman the phrenological reading that characterized him as the ideal poet, on his way to becoming the poet of American democracy.[27]

In 1850, after leaving the editorship of the *Brooklyn Daily Eagle*, Whitman began to work as a bookseller. Prominent among the books he peddled were titles by Fowlers and Wells. Reading the books, internalizing the names, locations, and functions of phrenological organs, feeling the promise of improvement at the center of the science, Whitman also began writing poetry in earnest, leading to the publication of *Leaves of Grass* in 1855. As he studied the science and experimented with verse, several references to phrenology began creeping into his writing, such as when he included the "phrenologist" on the list of "lawgivers to poets." Similarly, in the poem "By Blue Ontario's Shore," he posed several questions for those who would endeavor to know the nation:

> Who are you indeed who would talk or sing to America?
> Have you studied out the land, its idioms and men?
> Have you learn'd the physiology, phrenology, politics,
>   geography, pride, freedom, friendship of the land?
>   Its substratums and objects?

What's more, Whitman routinely used the names of phrenological organs in his attempt to capture American character, including Amativeness, Adhesiveness, Combativeness, and similar terms. He spoke of brains divided into different mental faculties, and he described joy in a way that seemed drawn from a phrenologically well-balanced mind.[28]

As Whitman embraced phrenology, the Fowlers took notice. When *Leaves of Grass* first appeared in 1855, the Phrenological

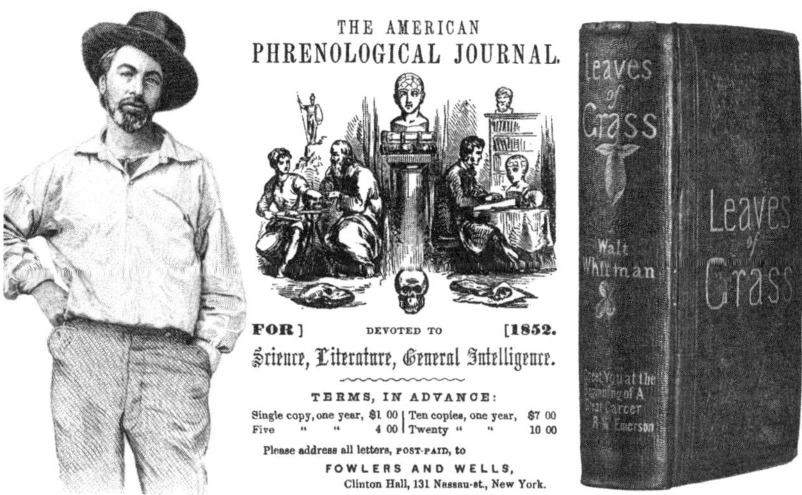

Walt Whitman (1819–1892), around age thirty-five when he peddled Fowlers and Wells material and published Leaves of Grass *with the firm*

Cabinet was one of only two booksellers in New York to carry it. They even publicized its availability in the *New York Tribune* and *The American Phrenological Journal.* Despite this publicity, the first edition was a major flop. Yet the Fowlers believed in Whitman enough to give him work as a correspondent for their new periodical, *Life Illustrated*, which covered art, science, literature, politics, and life. Whitman's articles were brief but regular, appearing once or twice a month and covering a range of topics—everything from the latest opera to the newly opened Egyptian Museum on Broadway.[29]

Even though the first edition of *Leaves of Grass* struggled for sales, *The Criterion* in New York got a copy, concluded that it was sinful and obscene, and decided to publish a vitriolic takedown that called for the book to be pulled from store shelves. Not one to shy away from a fight, Whitman showed the takedown to his friends at Fowlers and Wells, and *Life Illustrated* quickly offered a rebuttal, noting that the great Emerson himself, with his own "refined and delicate nature," cherished the book. If delicate Emerson was not offended by it, no true American should be. Fowlers and Wells continued to publicize and sell the book at their Cabinet.

They also saw an opportunity. Samuel Wells concocted a plan

to bring the poet of democracy formally into their operation, but to do so with what he hoped would be minimal reputational risk to the firm. Fowlers and Wells agreed to publish a second, expanded edition of *Leaves of Grass* with all the bells and whistles at their disposal. A beautiful object highly prized even today, the book was expertly bound in supple leather with gold stamping on the spine. The words in gold were in fact from Emerson's message to Whitman: "I Greet You at the Beginning of a Great Career." Despite the beautiful object they created, Fowlers and Wells did not put their name on it, publishing the book anonymously so they could make money without sullying the reputation of the firm, should charges of obscenity continue.[30]

In the lead-up to publication of the second edition, which appeared in September 1856, the firm put their publicity machine into high gear. They printed several puff pieces about the first edition in order to drum up interest in what they would soon have on sale. *Life Illustrated*, for instance, praised Whitman as "more a democrat than any man we ever met," then continued:

> He believes in American principles, American character, American tendencies, the "American Era," to a degree that renders his belief an originality. When he explains, in "Leaves of Grass," "By God! I will accept nothing which all cannot have their counterpart of on the same terms," he expresses the very soul of democracy.[31]

Several other puff pieces followed this one, as did advertisements in all their publications, as well as in the *Tribune* and *The New York Times*. The Fowlers even allowed Whitman to write a lengthy, unsigned, multicolumn review of his own book, which they published in *The American Phrenological Journal*. The review was carefully constructed to position Whitman as America's democratic voice:

> Not a borrower from other lands, but a prodigal user of his own land is Walt Whitman. Not the refined life of the drawing-room—not dancing and polish and gentility, but

> some powerful uneducated person, and some harsh identity of sound, and all wild free forms, are grateful to him.[32]

Yet just as Wells had feared, controversy continued to plague the book. Critics again called it obscene, and Wells soon felt the heat. Worried of what the controversy might do to the firm, he concluded that Fowlers and Wells could no longer continue offering it. For his part, Whitman had grown sour on the partnership too. In June 1857 he wrote to a friend about the unhappy arrangement:

> Fowler and Wells are bad persons for me. They retard my book very much. It is worse than ever. I wish now to bring out a third edition, . . . and shall endeavor to make an arrangement with some publisher to take the plates for F. & W. and make the additions needed, and so bring out the third edition. F. & W. are very willing to give up the plates—they want the thing off their hands.

For at least another year, the Fowlers continued to list the book in their offerings, but it was the swan song of Whitman's relationship with the firm. His time as a correspondent for *Life Illustrated* and as an author with Fowlers and Wells had come to an end.[33]

Yet his love of phrenology persisted. However much he may have soured on the publishing arm of the Fowler operation, he remained a devotee of the new science, continuing to reprint his phrenological chart from 1849 as evidence of his poetic mind. He also continued to weave phrenological terms and ideas throughout his writing. Even at the end of his life, at a time when most Americans had dismissed phrenology as foolish and wrong, Whitman claimed its insights. A few years before his death, he explained to a friend in June 1888 that he had been cautious for most of his adult life because of Lorenzo's phrenological reading from 1849. As he put it,

> Thirty years ago or more a circle of *celebres* in phrenology gave my head a public dissection in a hall—for one

> point, marked my caution very high—seven and over. Their seven was backed by my experience with myself. I live even today most conservatively—avoiding things that would be sure to be fatal to me. I know what Holmes said about phrenology—that you might as easily tell how much money is in a safe feeling the knob on the door as tell how much brain a man has by feeling the bumps on his head: and I guess most of my friends distrust it—but then you see I am very old fashioned—I probably have not got by the phrenology stage yet.

The nation had moved on, but the poet of American democracy had not yet left the phrenology stage. Old-fashioned, he still fondly remembered the days when the new science commanded the nation's attention and pointed him toward a life of poetic genius. Improvement and hope—they were the watchwords of phrenology and central to Whitman's view of life, regardless of what the scientists said.[34]

CHAPTER SEVEN

WHAT'S SEX GOT TO DO WITH IT?

THE FILTH, CRIME, AND VICE OF NEW YORK'S FIVE POINTS neighborhood did little to stem the tide of visitors to 128 Duane Street. Just the opposite, in fact—people went to the house *because of* the debauchery.

Knocking on the heavy wooden door, visitors were greeted by a woman made up to perfection and adorned with all the niceties of fashion. Her name was Mary Berry, but she often called herself "Duchesse de Berri," a nod to Madame du Barry, who had scandalized the French court in the late eighteenth century because of her low birth and licentious behavior. The house on Duane Street was not part of some royal court, at least not beyond Five Points. But it was a palace of earthly delights that made its proprietress rich.

Playing up the royalty schtick to great effect, Duchesse de Berri gave her business an appearance of gentility to accompany the flowing liquor, rigged dice games, and unfettered sexual escapades—at least according to what the guest could afford. While she saw to the pleasures of the guests, her husband, Frank Berry, usually robbed them, lifting whatever he could from their clothing while they were distracted. Frank often got away with the robberies, but even when caught, he knew how to handle himself in a fight. If Frank needed to bash a skull, he would. If, for some reason, the customer called the cops and Frank had to spend the night in jail, the money earned through robbery and women would more than cover the fine. And for every angry customer who felt swindled, there were five more at the door ready for a romp with Mary's girls.[1]

Five Points, *an 1827 illustration of the New York neighborhood considered the most depraved on the planet*

The brothel at 128 Duane Street was, in fact, just one of over a dozen other brothels in the neighborhood, which was also brimming with saloons, gambling dens, assaults, murders, robberies, even public sex. Spanning the few blocks between what is now City Hall Park and Columbus Park in Lower Manhattan, Five Points was known at the time as the most depraved neighborhood in the Western Hemisphere. Women lined the streets bare-chested and hollered to men, even those whose wives were on their arms, while trading insults with the other bare-chested women competing for customers. There were even guidebooks to brothels sold at newsstands, the most tongue-in-cheek of which was Butt Ender's *Prostitution Exposed; or, a Moral Reform Directory*. For the sake of public morals, and so visitors knew which houses to avoid—wink, wink—the book listed the address of every brothel in the city, along with the names, ages, and demographics of the girls working there.[2]

Butt Ender had sin and profit on the mind when he published his directory, but earnest reformers visited Five Points with salvation on the mind. Members of the New York Female Moral Reform Society regularly stormed brothels to break up the transactions, rescue the women, and shame the men who violated female innocence and the sanctity of the family. For every john and working girl they removed from the neighborhood, two more showed up to take their place. Ministers also visited the neighborhood to save souls. Dressed in a stately top hat and overcoat, Reverend John McDowell marched into brothels armed with Bibles and reform tracts, which he distributed to the women and men in attendance before saying a prayer for their redemption.

Then there were the public health officials who tried to curb the social ills of the neighborhood. In 1858, physician William Sanger published an official report on prostitution in New York that featured survey responses from over two thousand working girls. Zealous and eager to make a difference, Sanger laid bare the "syphilitic taint" of Five Points and the "moral pestilence which creeps insidiously into the privacy of the domestic circle, and draws thence the myriads of its victims, and which saps the foundation of that holy confidence, the first, the most beautiful attraction of home." Sanger's campaign, like the other reform efforts, did little to abate the problems of Five Points.[3]

Perhaps there was another way to address those problems. Perhaps there was a way to treat what happened in Five Points not as the cause of social ills but as the effects of something deeper, something more basic, something rooted in human nature itself. At least that's what the Fowlers thought as they passed the neighborhood, which happened to be only a few blocks from the Phrenological Cabinet. Working at the Cabinet day in, day out, they had a front-row seat to the roiling pestilence of prostitution, gambling, drunkenness, robbery, and murder. They also watched as moral reformers stormed the neighborhood, preached of repentance, then departed, only to have the revelry of sin continue unabated.

The more the Fowlers watched the cycle of sin and reform, the more convinced they became that phrenology not only could intervene but must intervene. It was a science of sex, among other things, able to address sexual sins as well as reproduction, parenting, family, and flourishing. But more than that, phrenology's approach to sex opened the door to something else crucial—namely, the equality of the sexes. In the Fowlers' work, sex was linked indelibly with gender progress and the rights of women. All of it stemmed from a potentially troublesome organ at the base of the head known as Amativeness.

AMATIVENESS GOVERNED SEX. As the Fowlers' *Illustrated Self-Instructor* pamphlet, given out with most phrenological readings, explained, it encompassed "conjugal love; attachment to the opposite sex; desire to love, to be loved, and marry; adapted to perpetuate the race." Though necessary for the species, it was easily perverted in both men and women. And to correct the sexual sins of modern society and to advance the place of women in America, the Fowlers knew they needed to tackle Amativeness head on. Of course, discussing sex openly and directly, even in the pages of a scientific treatise, was taboo, if not off-limits, for many Americans in the decades before the Civil War. So the siblings had to tread delicately.[4]

Which they didn't. "Parents and teachers," Lorenzo announced in his 1846 treatise *Marriage: Its History and Ceremonies*, "must lay aside their false delicacy, and teach their children . . . a more even balance of mind." This was a call for sex education at home and school, something rarely done at the time. Yet if children didn't learn about it properly from teachers and parents, they would learn about it in all the wrong ways. They would learn about it, Lorenzo warned, by attending "parties of pleasure, clubs, carousels, balls," where the "free use of *stimulating* food and drink, in connection with the exercise of social feelings" leads to debauchery. They would also learn about it by "reading works of romance written by persons of morbid feelings, sickly sentiments, and extravagant hopes," and by "attending theaters and other similar places of amusement, whose principal attractions are unnatural and far-fetched representations of scenes overloaded with 'love.'" Parents and teachers—and phrenologists, of course—needed to step in now and teach about sex so the youth could avoid the sins of their elders.[5]

If Lorenzo sounded like a Victorian moralist by pointing to

Love very Large.

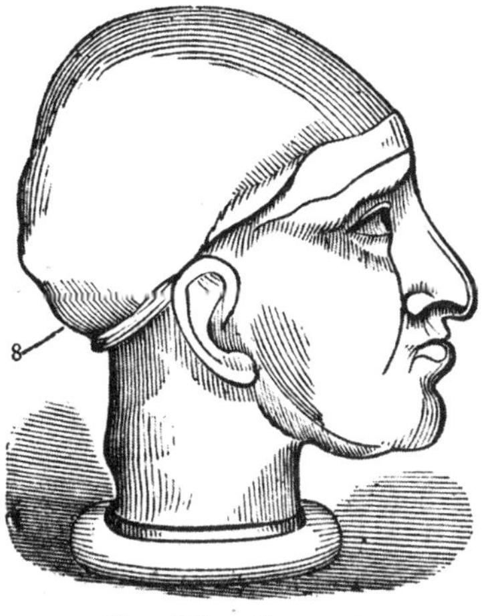

Fig. 507. — Gottfried.

Love Small.

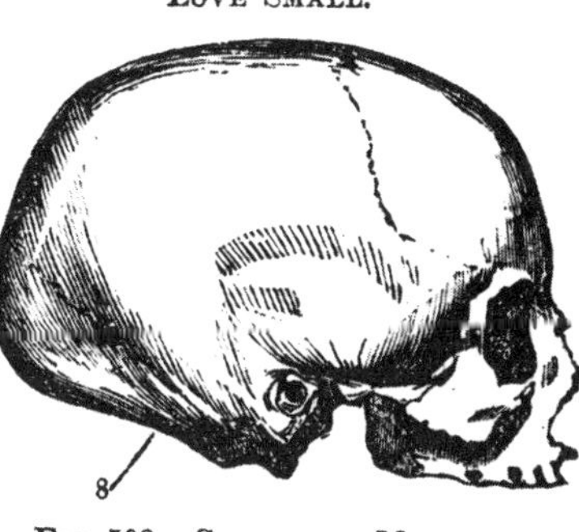

Fig. 508.—Skull of a Maiden at sixty, who died in the poor-house, was taken to the dissecting-room, and found to be a virgin; obviously from sexual indifference. This organ is scarcely perceptible.

Comparative figures used in many Fowler and Wells publications showing overdeveloped and underdeveloped Amativeness

parties, novels, and theaters as sources of sin and corruption, that's because he was. Yet he wasn't the same kind of reformer as those who burst into brothels and prayed for souls. He knew there was no quick fix to the problems of sex that plagued the nation. The solution required years, even generations, of systematic inquiry that would direct people away from the likes of Five Points before they were tempted by it.

Specifically, the solution to the problems of sex in modern society was the proper adaptation of two minds to each other—an important point about gender equality that pervaded Lorenzo's thinking. In his view, adapting two minds to each other was a matter not just of marriage but also of marrying people with constitutions that would allow them to thrive together—so that young women wouldn't end up in brothels, and young men wouldn't be prowling for sex. Men and women needed to adapt their Amativeness to each other and meet in the middle of sexual desirc. As Lorenzo explained,

> From my extensive observations and knowledge gained by fifteen years travel in all parts of the country, and

> becoming acquainted with families from various parts of the world, I have at times almost arrived at the conclusion that one-half, if not more, of all difficulties existing between husbands and wives, and premature deaths, are produced by want of proper adaptation to each other in this organ.[6]

To illustrate the point, Lorenzo spent much of *Marriage* giving examples of people whose Amativeness in particular and phrenological profile in general led them to sinful, ruinous places. He told stories of mothers locking up their children so they could go out on the town and find sex, of women whose unrestrained passions led them to murder and mayhem, of young ladies who devoted their leisure time to amusements and ended up as prostitutes, of babies who were thrown to the garbage because their mothers had no drive to care for them, of men who preferred the company of their friends over their family and ended up at pleasure houses, of drunkards who ceaselessly satisfied their craving for liquor and destroyed their families in the process.

Overcoming these evils required developing proper habits of mind and body, treating marriage as a rational, scientific decision as well as a decision of feeling, and raising children in a way that fostered balanced mental faculties. Simply put, improving sex in society meant practicing phrenology. "Study well your own character, and that of the one whom you select as your partner," Lorenzo declaimed at the end of his treatise, and you will create a heaven on earth for you, your family, your community, and your nation.

Although Lorenzo's comments on sex and society seem quaint these days, his treatise included some impressively progressive ideas for the pre–Civil War decades. This was especially true in his remarks on prostitution. Lorenzo argued that prostitution was a sin less because of sexual immorality than because of the way it pushed women into the gutter of society. "Where is the fallen, ruined victim's redress? Where, even in enlightened America, is to be found a reparation for her grievous wrong? She is cast out from the pale of society, despised by her own sex, and has a mark set upon her as ineffable as that which was branded upon the forehead of Cain." At the same time, the man who pushed her into prostitution, this "infamous villain," is "allowed to walk forth in society with all the apparent self-complacency of a knight-errant and is even caressed by kindred spirits on account of the cool and skillful execution of his crime." Same goes for the john who uses women for sex. Expressing his "abhorrence and detestation of such a monster, and of those who would teach him to become skillful in his piratical career," Lorenzo put the blame for the evils of prostitution squarely on men, not on the supposed sexual perversion of women. Inequity and injustice in the way America treated men and women—that was at the root of the brothels of Five Points.[7]

LORENZO APPROACHED AMATIVENESS with a sense of justice and a conviction that discussions of sex needed to happen regularly in America. Orson took a somewhat different course, firing a cannonball into the pillars of civil society that he hoped would destroy the structures of decorum. Unfortunately, several decades later, the cannonball he fired would explode the foundations of his own career, as he ended up in a sex scandal that made national headlines and severed ties with his siblings.[8]

Back in 1844, Orson was in his mid-thirties and still full of cocksure energy that phrenology would save the world, even when it came to sex. His first big attempt to tackle the issue came as a seventy-page pamphlet on the nature, perversions, and effects of Amativeness. Pointing to the suffering that results from the improper development of the organ, he bemoaned the nation's preachers for their "utterly unjustifiable" omission of treating sexual sin in their sermons. He denounced members of the medical profession for staying "mostly silent" on the health consequences of licentious behavior. He lambasted lawyers for the fees they charged to handle the legal consequences of sexual impropriety, which resulted in more money "than from any other crime perpetuated by our fallen nature." Standing before human ruination in places like Five Points, he lamented, "the moralist is silent, and the philanthropist is dumb." What the world needed was a fearless phrenologist.[9]

In the first fifteen pages of the pamphlet, Orson made clear to readers that he was unafraid to speak boldly where most chose to remain silent. Decrying the "great maelstrom of the devouring pit" of hell that sexual sin created in society, he pointed readers to "the number of seductions, of abortions, and of illegitimates, which annually disgrace our age!" He lambasted the rich and powerful for their evil appetites. "VIRGINITY SOLD AT A PRICE," he wrote in all caps. "Wall-street brokers actually speculate in maidens! Mothers sell their own yet unpolluted daughters to beastly sensualists!" Indeed, it was greed that pushed people to do unthinkable harm to others, resulting in disease, pain, and death.[10]

After Orson's full-throated denunciations of the greedy and powerful, he turned directly to the subject that was, at the time, the most unspoken sexual sin of them all—masturbation. Self-abuse, self-pollution, private fornication, the filthy practice, the accursed plague—he had wonderful names for talking about an action that, he believed, threatened the very foundations of society. It was the root of ruin for countless boys, he explained, making some insane, leading others to brothels, and putting still others in the morgue. He estimated, based on no real evidence, that "thousands on thousands die annually from this one cause!"[11]

But—and here Orson really exploded the subject matter—the evils of masturbation were not limited to men. Female self-gratification, he explained over several pages, was horrifyingly common among young girls as well as aged prostitutes. Most women in brothels, he diagnosed, actually started as chronic masturbators. He even shared a story with his readers about a time he had to speak some hard truths to a mother and her daughter. After a quick phrenological examination of the girl, he pulled the mother aside to disclose his suspicions that the girl was headed down a dangerous road of self-satisfaction. The mother was aghast. There was no way, she told Orson, her little princess was partaking in such a vile activity. But the more he chatted with the mother, the more she realized that her daughter had been spending lots of time with an older girl in the neighborhood who was known to be a chronic masturbator. Then the mother recalled a nearby house where a number of factory girls lived (and apparently a bunch of factory girls living together meant unabated self-gratification). The mother also recalled a girl in her neighborhood who had recently died of masturbation. It all came into focus for the woman, who hopefully had enough time to save her daughter from an early death in Five Points.[12]

Orson wrote about masturbation with an even stronger sense of Victorian moralizing than Lorenzo wrote about Amativeness. And yet even in his conservatism he was boldly progressive for the time. Open discussion of prostitution, male and female masturbation, the sale of virginity, and effects of sexual disease was well beyond most published discourse of the era. But Orson was not afraid of offending sensibilities; he was interested above all in justice and progress, especially for women. One of the biggest problems with sexual perversion, he told readers, was the way it "deteriorates woman in the estimation of man." Overly active Amativeness in men made them treat women as sexual objects rather than as thinking, feeling, whole human beings. A man who feeds his carnal appetites, Orson wrote, "has been mainly conversant with woman as a *sexual thing*, and not as a pure, refined, affectionate being. Her sexuality mainly is what he has noticed, and this he detests in himself, and therefore in her."[13]

Unsurprisingly, the solution to these problems was enacting the principles of phrenology. To Orson, this meant abstaining from masturbation, avoiding intoxicating beverages, soaking oneself in cold water when needed, keeping busy to resist temptation, and finding the right marriage partner. None of these were terribly novel suggestions, but they emerged from a fearless approach to the problems of sex and a keen interest in leveling the social standing of men and women.

In 1848, Orson followed his earlier work with a book on reproduction titled *Maternity; or, the Bearing and Nursing of Children, Including Female Education and Beauty*. Insisting that he wanted to bring supposedly "improper and injurious" subjects into "the sunlight of Phrenology, Physiology, and Magnetism," the book was sex education for families—for expecting mothers in particular. It covered such topics as "the female secretion" during pregnancy; proper care for embryos; regular bowel movements; the need for good food, fresh air, and exercise; and the duty of husbands to care for pregnant wives. Always the storyteller, he also relayed several anecdotes involving maternity from his phrenological encounters. In one encounter, Orson examined a woman whose vitality was not sufficient to endure childbirth. His counsel was for her never to have kids—a troubling piece of advice to a woman who happened to be three months pregnant. Orson heard an earful about that the next day when the woman's friend returned to his office to scold him.[14]

The most fascinating maternity-related anecdotes he told involved people with deformities whose mothers had experienced some kind of trauma while pregnant. The particular nature of the trauma, he claimed, ended up affecting the fetus in remarkable ways. One anecdote involved a woman who had an extra thumb on one hand, producing what seemed like a lobster claw. This "lobster mark" was the result of her pregnant mother having a lobster stolen from her before she could eat it for dinner. Another example was of a man with a mark on his leg that resembled a mouse. Apparently his mother had been terrified of a mouse scurrying around the bedroom during her pregnancy. Yet another example was of a child born with a stump instead of a proper thumb, almost as though the thumb had been amputated. The mother had seen her husband's thumb cut off while she was pregnant, and the horror of that incident was transferred to the fetus. A more horrific example involved a woman in Vermont who visited a local menagerie of animals and became enthralled with the exotic specimens she saw there. A few months later she gave birth "to a monster, some parts of which resembled one wild animal, and other parts other animals. It died soon after."[15]

Orson offered these anecdotes as warnings to expectant mothers about the way trauma can affect gestation. Because of the ethereal fluid permeating creation, one magnetic charge could jump from the environment to the mother to the fetus, with potentially horrific effects. Of course, he was wrong about this. There is no magnetic, ethereal reality pervading everything, and a child born with a lobster-looking hand did not receive the hand because her mother had a lobster dinner stolen. But his concern for pregnancy, childbirth, and women's health—coupled with the practical tips he offered—connected phrenology directly to a female audience.

Same was true of his thoughts on women's clothing. *Maternity* spent well over forty pages on the subject of what women should and should not put on their body. He wrote an entire section on "cotton and plaited breastworks," or pieces of clothing that made women seem to have larger breasts than they really did. And he saved special ridicule for "tight lacing," which he called "the most accursed of all fashions." "She hangs all this extra-clothing upon her hips and bowels, and this of necessity presses down her female organs, gradually displaces and disorders them, and thus weakens and diseases that specific function, the perfection of which constitutes female perfection, and the impairment of which blights the very essence of the female nature." In his typical hyperbole, Orson even went so far as to deem "licentiousness, in all its forms and degrees," nothing compared to the evils of clothing that harmed a woman's maternal function.[16]

These admonitions about women and their proper role in the world (childbearing) are problematic today. Yet as with so many of the Fowlers' writings, it's easy to pass over the progressive elements within. Denouncing tight-lacing was a way of encouraging women to throw aside the impositions of modern fashion, to break free from domestic confines, and to get out into the world to develop their minds and bodies. Women were best served, he wrote, by "rendering yourself healthy." That meant "an abundance of vitality, air, exercise, and good digestion." It also meant exercising the intellect: "Let the bearing mother study."[17]

Although it was strange for Orson to hand down instructions on women's clothing, he wrote as a physician and progressive reformer intent on liberation. And he was just one of many doing so. In the mid-nineteenth century, dress reform was a leading issue for women's rights activists, including Amelia Bloomer, who spoke and wrote extensively about the health hazards of dresses and other accoutrements. She argued that women should dress in a way that improves their health and promotes their equal footing in civil society. Orson argued the same thing.[18]

AS THE FOWLERS attuned practical phrenology to the needs, interests, and rights of women, many female reformers took note and began to embrace the science as integral to their cause. Perhaps the earliest suffrage leader to embrace phrenology was Lucretia Mott. Mott attended the lectures of George Combe, Europe's leading phrenologist, when Combe traveled to the United States at the end of the 1830s. Combe even examined Mott's head and found her Causality to be significant and her Veneration to be large, perfectly adapted to abolitionism, which Mott deemed to be the true work of God. She was so smitten with phrenology that she wrote a letter to Combe suggesting that perhaps the "truths of Phrenology" could reform churches and schools and "substitute for the dogmas and hidden mysteries there instilled." In 1853, Orson published his own examination of Mott's head alongside the earlier examination by Combe, just to show how closely the readings aligned.[19]

Several other women's rights activists took up the cause of phrenology as well. One was Abby Kelley Foster, who regularly referenced the new science in her advocacy. She even followed Orson's advice about marriage when she was navigating her own romantic possibilities as a young woman. Another phrenology proponent was the writer, editor, and activist Margaret Fuller, who was working as a schoolteacher in Rhode Island when Orson came to town. Excited by the prospects of phrenology, she brought Orson to her class, asked him to speak to her students about the science, and had him examine their heads. When he was done, she sat for her own examination, although the report from it has been lost to history.[20]

Several years later, both Elizabeth Cady Stanton and Susan B. Anthony became devout phrenologists. On January 10, 1853, in

Seneca Falls, Stanton let Lorenzo feel her head and pronounce her character. Praising her large social and intellectual organs, he laid out how she was brilliantly suited to the path ahead for suffragists: "You have remarkable Firmness and perseverance when you have once laid out your plans and taken your position. You are not to be driven from your purposes or prevented from consuming your designs." He also noted her sizable intellect, which "lies chiefly in your disposition to reason" and in "your faculty to originate thoughts and follow your own ideas." After the exam, Stanton wrote to her father that the reading "often hits the nail on the head—I really did not mean to make a phrenological comparison—in a rather striking way."[21]

A month later, Susan B. Anthony popped into the Phrenological Cabinet in New York City to let Nelson Sizer feel her head, although she asked him to wear a blindfold while he did it, hoping to test the validity of the science. Supposedly unaware that the sitter before him was Anthony, Sizer developed a strikingly accurate description of her character. Noting her "intensity of emotion and thought," which makes her mind "terse, sharp, spicy, and clear," he praised her unflinching advocacy: You "unfurl your banner, take your position and give fair warning of the course you intend to pursue." He went on to laud her intellectual and reasoning faculties as well as her ability to judge character and to speak with "accuracy and definiteness," perfect characterizations of her suffrage speeches. So taken was Anthony by the reading that thirty years later she included it in her official biography.[22]

Stanton, Anthony, and many other women's rights leaders received their phrenological examinations at the hands of the Fowlers, not just because of the influence Orson and Lorenzo had achieved, but also because of the importance of the Fowler women, Lydia Fowler and Charlotte Fowler Wells. True pioneers in their respective professions, they seamlessly integrated the science with women's rights and achieved national influence on par with, if not beyond, the Fowler men.

FROM HER VANTAGE point at the front of the auditorium, Lydia Fowler could see the full, awful ordeal. This, she knew, was what woman suffrage was up against. This was the ignorance—here in New York of all places!—they would have to overcome nationwide if women were to receive the vote. The ridiculousness of the situation was compounded by the fact that each of the rabble-rousers paid fifty cents to be there; they were financially supporting the women's rights movement at the same time they were trying to derail it.

The mob was already riled up from earlier in the week when they disrupted an abolitionist meeting. On Sunday, September 4, 1853, the New York Anti-Slavery Society assembled in Metropolitan Hall to rally support after a series of legal actions had knocked the antislavery cause on its heels. In attendance were all the big names in the movement, including William Lloyd Garrison, Lucy Stone, and Lucretia Mott. But the more these leaders tried to make their case for abolition, the more they were hit with waves of hisses and jeers from the rowdy crowd. The noise eventually became so great that they had to end the meeting early.

Also on that Sunday, Antoinette Brown, the first woman ordained as a mainstream Protestant minister in the United States, delivered a sermon to five thousand people. The service went relatively smoothly, but the conservative pulpit and press reviled the idea of a female minister and deemed Brown's preaching yet another instance of America's moral corruption. "Thus we see paraded before us," wrote one angry columnist, "all the live humbugs of the land."[23]

Given the disruptions of earlier in the week, it was really no surprise that the Woman's Rights Convention of September 6 and 7, held at the Broadway Tabernacle, ran into trouble of its own. In point of fact, many of the reformers who had been at the Anti-Slavery Society meeting and had listened to Brown's sermon were also present for the convention. One by one they rose to address the audience, and one by one they were battered by shouts, hisses, and jeers from a small but vocal part of the crowd. The mob made clear, according to Stanton, the extent of "public sentiment woman was then combatting." While no fights broke out, the scene demanded patience

Lydia Fowler (1823–1879), a proud phrenologist and the second woman to earn a medical degree in the United States

from the speakers, one of whom was the renowned former slave and champion of justice Sojourner Truth. The mob tried to drown out Truth's speech, but she stood there with stoic strength, waiting for the young men to tire themselves out. They eventually did, and she was able to deliver her remarks. But tensions only escalated after she finished, and by the end of the convention, supporters of women's rights were exchanging shouts with the troublesome mob. When nothing could be heard over the commotion, the convention was forced to come to an end.[24]

That was fortunate for Lydia Fowler. As secretary of the convention, she was charged with taking minutes and seeing to an orderly process. But the mob made that impossible, and she could only look on and wonder. Why did they so oppose a peaceable assembly dedicated to human rights? Had she been able to examine their heads, she likely would have found their Combativeness and Destructiveness large. Given that they were all blasted drunk, she probably would have found their Alimentiveness, or love of food and drink, large as well. It's likely that their Conscientiousness, or sense of equality and justice, was small. The same was probably true of their Ideality, or sense of refinement, taste, and purity. Theirs were not the heads of wise leaders or temperate young men.

By the time of the 1853 Woman's Rights Convention, also known as the "mob convention," Lydia Fowler had been married to Lorenzo for almost ten years. In that time she had come to know, practice, and love phrenology, especially what it meant for women. And in many ways she had already surpassed Lorenzo's legacy, given her accomplishments in the medical field.

Born in 1822 on Nantucket, Massachusetts, Lydia Folger, her maiden name, came from a prominent family. Her father, Gideon Folger, was a merchant and descendant of Benjamin Franklin, something that people regularly mentioned when the subject of Lydia's lineage arose. Her uncle, Walter Folger, was a preeminent astronomer and politician, the most famous man then living on Nantucket. Lydia also had a famous cousin, Lucretia Mott, the prominent abolitionist and suffragist who, early on, had taken to phrenology.

Lydia fit right in among such learned, influential figures. Precocious from an early age, she hungered for new knowledge as a student at the Lancasterian School in Nantucket, where she helped teach the younger students and, alongside her sister, volunteered to assist the principal. When she was sixteen years old, she matriculated to Wheaton Female Seminary in Massachusetts, which she attended from 1838 to 1839. She returned to the school as a teacher from 1842 to 1844. But that year, her life changed forever via the hands-on work of a handsome, bright-eyed lecturer.

Lorenzo traveled to Nantucket in March 1844 to spread the gospel of phrenology and to examine some heads. One of the heads he examined was that of Lydia's well-known uncle, Walter Folger. But it was Lydia who really caught his eye. For some unknown reason, Lorenzo did not examine her during that first visit, but he came up with an excuse to return to Nantucket in April and to reconnect with the beautiful, intelligent young woman.

Standing behind her, pheromones swirling in the air, Lorenzo ran his nervous fingers through her luxurious hair. His hands worked slowly across her impressive cranium, touching her head perhaps longer than needed. By the end of the exam, Lorenzo knew that he was feeling someone special. Her body and brain were perfectly adapted to each other, and her scholarly aptitude practically leapt off her skull. "She is constantly taking lessons," he announced. "She learns something from everything she sees, hears, or reads. She observes and obeys the language of nature." After describing her character for several more minutes, he returned to the intellectual powers of his future bride: "She has a critical turn of mind, with sagacity, intuition, and perception of hidden motives, and her desire to become acquainted with recondite springs of action, and her disposition to form a like or dislike at first sight, are full qualities of mind."[25]

Lydia's ability to form an opinion at first sight certainly was true of the phrenologist feeling her head. She fell in love quickly with the man who respected her mind above all. Five months later, Lorenzo and Lydia married and moved to New York City to share a home with Charlotte and Samuel Wells.

As soon as she married Lorenzo, Lydia put her intellectual powers to use studying the new science, learning to examine heads, and helping out at the Phrenological Cabinet. She was especially interested in what phrenology meant for women and children. After a couple years, she began accompanying Lorenzo on his lecture tours and delivering talks of her own. While he lectured to men, she spoke in a separate room to women. As one newspaper summarized, Lydia's "interesting and useful lecture to the ladies showed that intellectual worth and moral goodness is not confined to one side of the house; and that woman possesses within her mind the elements, which, if properly improved, will do much to enlighten and benefit her race."[26]

If that were the extent of Lydia's contributions, her life would have been significant enough. But at the time she was studying, practicing, and speaking on phrenology, a controversy was brewing that would ultimately etch her name into the annals of history. At the end of 1847, Harriot Hunt tried to enroll in medical classes at Harvard. The dean of the medical school, Oliver Wendell Holmes Sr., was at first inclined to admit her, but pushback from the all-male student body prompted him to deny Hunt's matriculation. Soon a battle over women in medical schools erupted in the national press, including in *The American Phrenological Journal*. Indignant that Harvard would not admit Hunt, Orson published several articles denouncing the school's decision. "We know Harriet [*sic*] K. Hunt," he wrote. "She is a good woman and true—intelligent, an eminently successful practitioner, imbued with the right spirit, and entitled to a hearing. Persevere, good sister, till your rebukers shall be rebuked by the people, and reaction shall crown your laudable application with success."[27]

The public fight over female physicians gave Lydia an idea. In 1849, at twenty-seven years old, she joined seven other women and enrolled at the Central Medical College of Syracuse. To support her application, she enlisted the help of her famous cousin, Lucretia Mott, who argued for her admittance on the grounds of justice and progress. Lydia spent the year studying medicine in Syracuse, but financial turmoil, coupled with faculty in-fighting, forced the school to relocate to Rochester. So in 1850, Lydia went there to study, ultimately graduating in June. She was, in fact, the only woman of the eight who originally enrolled to receive her degree, which made her the second woman in the United States to earn an MD. Elizabeth Blackwell had earned her degree in 1849, but Blackwell was British by birth, so the proud phrenologist Lydia Fowler had the distinction of being the first American woman with an MD.[28]

When Central Medical College moved to Rochester, it established a "Female Department" and asked its star female pupil to run it. With the title "principal" of the Female Department, Lydia lectured on anatomy and physiology to other women at the school. Then in 1851 she was named Professor of Midwifery and Diseases of Women and Children, making her the first woman professor at a medical school in American history. But the position did not last long. The school continued to face financial problems and political in-fighting, and it formally shut down in 1852. With that, Lydia moved back to New York City to be with her husband and children and to practice medicine out of her own shop, where she lectured to women on "Anatomy and Physiology, illustrated with a skeleton, mannequin, drawings and models."[29]

As the first American woman to earn a medical degree, Lydia had important visibility in the cause of women's rights, a cause she wholeheartedly supported by attending suffrage and temperance conventions and working alongside the most famous activists in the nation. She was part of the Akron Woman's Rights Convention of 1851; an organizer, along with Susan B. Anthony, of a New York temperance convention in 1852; and the president of a mass meeting in New York in spring 1853, which was followed by a speaking tour across upstate New York. In the fall of that year, Lydia accompanied Elizabeth Cady Stanton on a tour across Wisconsin. Speaking on physiology and phrenology, Lydia took a stance against "the world's dread laugh" over any woman who should try to "accomplish some public good" and "go out into the world alone," as Stanton later recalled. Lydia's advocacy made national headlines in 1853 when she, Lucy Stone, Susan B. Anthony, and Amelia Bloomer dressed like men to present their demands at a temperance convention. The group of "very gentlemanly looking ladies," wrote the *Vermont Journal*, was led by Lydia Fowler, who "was habited in a short upper garment of silk that was neither a coat, nor exactly a petticoat, and a pair of loose pantaloons. The whole troop of belles were in bifurcations, and made quite a grotesque appearance."[30]

By the end of the 1850s, Lydia Fowler was not only the nation's most recognized female phrenologist but a trailblazing leader in the movement to secure women's rights. For her, phrenology and social progress for women were inseparable.

WHILE LYDIA FOWLER was leading the charge publicly for phrenology and rights, Charlotte Fowler Wells was working strategically behind the scenes. And though her contributions were relatively quiet prior to the Civil War, she emerged in postwar decades a nationally recognized business leader.

Charlotte and her husband, Samuel, controlled the publishing arm of the Fowler machine, deciding everything that appeared in the firm's output. In their role, they oversaw a redesign and expansion of *The American Phrenological Journal*, which included increased coverage of more general news, including politics, global affairs, agriculture, business, new inventions, and more. Women's rights, suffrage, and temperance were also regularly featured in the journal. All the major conventions of the day appeared in its pages, including the Worcester and Akron conventions of 1851; the West Chester and Syracuse conventions of 1852; the New York City, Albany, and Rochester conventions of 1853; the Albany convention of 1854; and more. Coverage of these gatherings involved not just notices but lengthy reports of proceedings, committees, and resolutions as well. It also publicized writings, speeches, and lectures by the likes of Amelia Bloomer, Elizabeth Oakes Smith, Harriot Hunt, Lucy Stone, Paulina Davis, Frances Gage, and many others.

At the same time Fowlers and Wells increased coverage of women's rights in its flagship periodical, the firm expanded its book list to include new titles on the movement. There were full-throated calls for rights and suffrage: *Woman, Her Education and Influence* by Marion Kirkland Reid; *Woman and Her Needs* by Elizabeth Oakes Smith; *Woman and Her Wishes* by Thomas Wentworth Higginson; *Hints Toward Reforms* by Horace Greeley; *A Few Thoughts on the Powers and Duties of Woman: Two Lectures* by Horace Mann; and *Woman's Rights Commensurate with Her Capacities and Obligations*, a collection of women's rights tracts that the firm bound and shipped for the low price of fifteen cents. There were also domestic-leaning publications for women, including *Hints on Dress and Beauty* by Elizabeth Oakes Smith; *Intemperance and Divorce; or, the Duty of the Drunkard's Wife* by Clarina Irene Howard Nichols; and *Biographies of Good Wives* by Lydia Maria Child. More generalist books by women authors included *The History of the Condition of Women in Various Ages and Nations* by Lydia Maria Child and *Literature and*

Charlotte Fowler Wells, tireless reformer and prominent figure in the fight for women's rights

Art by Margaret Fuller. The firm also published a novelesque exposé of women's healthcare called *Delia's Doctors; or, a Glance Behind the Scenes* by Hannah Creamer.

Yet the most significant Fowlers and Wells publication in the realm of women's rights came in the mid-1870s, when Susan B. Anthony and Elizabeth Cady Stanton began gathering materials for a comprehensive history of the movement. Writing to Charlotte in 1876, Anthony mentioned that she would be visiting New York to look for a publisher of the history, and she believed Fowlers and Wells fit the bill. In her view, the firm "had made considerable profit out of publications on a controversial subject and were less timid than most publishers in this respect." What's more, the women of the Fowler family "had been firm supporters of woman suffrage for many years." What Anthony neglected to mention was that most major publishers had already declined to publish the book.[31]

No matter, at least for Charlotte. Readily agreeing to take on the project, Fowlers and Wells became the publisher of the first two volumes of *History of Woman Suffrage*, edited by Elizabeth Cady Stanton, Susan B. Anthony, and Matilda Joslyn Gage. The book was dedicated to a number of influential women, including Lydia Fowler, and the authors saved a special shout-out for Charlotte herself, calling her "the efficient coadjutor of her brothers and husband for the last forty-two years in the management of *The Phrenological Journal* and Publishing House of Fowler & Wells in New York City."[32]

The book did well, at least by the standards one could expect for such a title. And that's where a fissure emerged. After the first two volumes were published in 1881 and 1882, Anthony realized that she could make more money by self-publishing the book and having it privately printed, so she bought the rights back from Charlotte and put out the rest of the history under her own name. Nonetheless, the firm had already made a prominent mark on the most significant book of women's history ever published in America.

It was all thanks to Charlotte, who remained committed to suffrage throughout her life. Less active than Lydia in public-facing roles, she nonetheless served in various capacities over the years, including the chair of the committee on publication for the 1850 Woman's Rights Convention in Worcester. More significantly, in 1863 she helped found the New York Medical College for Women and thereafter served as one of its trustees. In 1870, she served as president of the Southern (New York) Women's Bureau, and in 1879 she spoke on behalf of female journalists at the New York Press Convention.[33]

Yet all of these contributions were minor compared to her most important achievement—namely, running a massive publishing firm. Every few years beginning in the 1870s, newspapers and periodicals wrote stories about how hardworking, shrewd, and dedicated a businesswoman Charlotte was. A paper in Portland, Oregon, for instance, wrote a profile of her in 1877, noting that much praise is given to women who make their mark in literature or art, but women who make their mark in business, "doing the work as grandly and successfully as" men, are often overlooked. "Eminent among this class," the paper wrote, is Charlotte Fowler Wells, whose "business tact is without question the result of a life-long devotion to business. From the time that she joined her brothers in 1837, she has been an active worker in the intellectual and commercial field, and adds one more to the long list of names which prove that intellectual pursuits serve to prolong life, and to keep the faculties in the most healthful tone until advanced age."[34]

The following year a correspondent in New York wrote to the *Daily Milwaukee News* about Charlotte's significance and influence, praising "the high moral and intellectual standard" of *The Phrenological Journal*, which was replete with "practical value" for the "thinking masses" and due to Charlotte's leadership. Indeed, the paper continued, while her brothers stole the spotlight with phrenology, she "was the oracle behind the scenes to whom they went for counsel and strength." The publishing firm has survived, the paper concluded, because Charlotte was able to direct it "through storms which have sunk so many more formidable enterprises." Other papers called her "the most indefatigable of workers, the embodiment of health" and "a venerable publisher" who is, in fact, "the oldest woman publisher living, as well as the pioneer woman in the phrenological field."[35]

As the pioneer women in phrenology, Charlotte and Lydia had seen what the science's program of improvement could do not just for individual women but for the collective place of women in American society. Their advocacy pushed phrenology in new directions, drew an increasing number of female reformers into the movement, and proved instrumental in helping a flourishing business transform into something much greater.

CHAPTER EIGHT

LIFE IN THE OCTAGON

BY THE EARLY 1850S, THE FOWLERS HAD CONQUERED AN expansive territory in the name of practical phrenology. They had planted a flag deep in the realm of science, purporting to unlock the secrets of nature—human and otherwise. They had planted a flag deep in the realm of entertainment, drawing together a dynamic confluence of lecturers, artists, performers, and celebrities. They had planted a flag deep in the realm of reform, connecting their science to women's rights, temperance, vegetarianism, mental-health care, education, abolitionism, and more.

There was also an impressive flag planted deep in the realm of business. Fowlers and Wells was one of the top publishing firms in the United States, and the Phrenological Cabinet was a world-renowned draw for the curious and the skeptical. Its free admission made it a must-visit attraction in New York, and its whimsical goods and services made it incredibly profitable. Once people were in the store and could feel the truth of humanity for themselves, they purchased books and journals, skulls and heads, instruments and manuals, and, of course, personal examinations.

With their phrenological flags flying proudly across American society, they knew it was time to expand even farther. In 1851, they opened a branch of the Phrenological Cabinet in Boston, or the "Metropolis of New England" as they called it. Located at 142 Washington Street, just two doors down from the famous Old South Church, the new office addressed "the want of such a depot" long felt among Bostonians, who, according to *The Boston Daily Bee*, were apparently clamoring for the Fowlers' publications "in such quantities as many have heretofore wanted."[1]

The opening of the new branch was accompanied by a coordinated marketing blitz. In addition to publicizing it widely in their own periodicals, the Fowlers took out ads in *The Boston Daily Post*, the *Daily Evening Transcript*, *The Liberator*, the *New England Farmer*, and the *Boston Atlas*, among other outlets. Then Orson and Lorenzo teamed up for a series of public lectures that would invigorate interest in the science. Every Monday, Tuesday, and Friday evening in November and December 1851, one of the brothers spoke at the Tremont Temple on topics related to phrenology and physiology. After listening raptly to Orson or Lorenzo, audience members could head to the Boston storefront for "Professional delineation of character, with complete verbal and written opinions, and advice touching health, habits, occupations, faults, self-perfection, management of children," and more.[2]

Orson and Lorenzo did not personally manage the Boston branch—they were too busy in New York for that—but they left it in good hands. David Butler, a former employee at the New York Cabinet, and C. J. Hambleton, another associate, oversaw the office and came rapidly "into favor with our New England friends," Orson explained to readers of *The American Phrenological Journal*. He continued: "With these honest, enterprising and intelligent men, we shall enlarge and extend our operations in New England, until the Reforms we advocate shall become as familiar in every family as 'household words.'" A few months after the branch opened, the *Boston Evening Transcript* made a special plug for readers to visit the office, noting, "Phrenology maintains its ground with so much tenacity that the probability is there that a good share of truth is in its principles."[3]

Boston was not the only new terrain for the expanding business. In spring of 1854, the Fowlers secured a storefront at 231 Arch Street in Philadelphia. This move took them back to the city where they had established the Phrenological Atheneum, an early trial of the format that would succeed wildly in New York. In a historic building along a redbrick sidewalk, right next to where Betsy Ross sewed the first American flag, the Philadelphia branch was "well stocked with all of the valuable and Reformatory Works published at the New-York establishment." As with the Boston branch, the launch of the Philadelphia branch came with a promotional blitz designed to capture the popular imagination of the city—advertisements in the local press, favorable coverage from editors, and a public lecture series. This time, Orson was the one who traveled to the city to speak on the science and to lure visitors to the store.[4]

The most important decision the Fowlers made with the Philadelphia branch was to send their second-in-command, Nelson Sizer, to manage the business. Sizer moved there from New York in early 1854 and, once the new office was up and running, delivered a lecture series of his own. His skill as an advocate of phrenology caught the attention of the editor of the *Philadelphia Daily Register*, who, like so many reporters before him, decided to put the science to a blind test. After letting the hubbub around the new storefront subside, the editor went to 231 Arch Street for a personal reading, giving Sizer no indication of who he was. Coming away with a full written report, the editor dashed home and set the character description before his family for them to judge its accuracy. The verdict? It was "a perfect daguerreotype of our character. Certain it is, that several very important points of our inner life and consciousness, which are unknown to the world, were most strikingly accurate in the deliberation. We cordially commend Mr. Sizer to our citizens as an accomplished examiner."[5]

Beyond the expansion to Boston and Philadelphia, there were important changes afoot in New York as well. Early in 1854, the Fowlers received word that the Mercantile Library Association would be moving out of Clinton Hall to a larger building near Astor Place. What's more, Clinton Hall itself would be demolished to make room for a newer, better structure made of marble. The Phrenological Cabinet would have to relocate. After a brief search of the city, the Fowlers found the perfect space at 308 Broadway, two blocks above city hall and right between Duane and Pearl Streets.

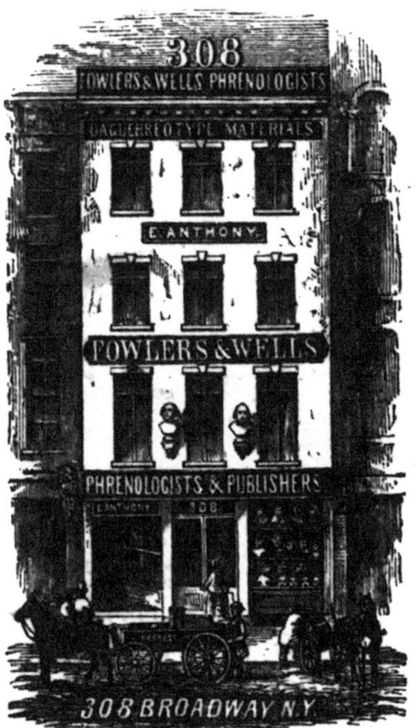

Exterior of the Fowlers' Phrenological Cabinet at 308 Broadway

It was, they insisted, a powerfully symbolic move for phrenology: "New York is an epitome of our country, and Broadway the heart's-core, seeds and all, of New York. We have procured for the Phrenological Cabinet one of the best stands in it, near its business center, where every promenade of Broadway must pass our door." The best thing about the space was its size. It was significantly bigger than the adjoining spaces they were renting in Clinton Hall—almost twice as large, the Fowlers announced—and it allowed them to display more of their collection, to perform more examinations, and to hawk more educational materials. They even opened a patent office in the Cabinet that assisted aspiring inventors with protecting their work—for a small fee and percentage of future revenue, of course.[6]

With the move to a larger space in New York and the establishment of the other branches, the Fowler firm was booming. The

Interior of the Fowlers' Phrenological Cabinet, originally printed in the New York Illustrated News, *1860*

siblings had spent twenty years traveling the country, explaining the science to countless Americans and examining hundreds of thousands of heads. They had created and released popular periodicals, best-selling books, and reform tracts that aimed to get the nation headed in the right direction. Their influence was felt in major cities and small towns, as well as in practically every major social movement. They had drawn together millions of people around a science that could improve individuals, families, communities, nations, and the world—and they had made a lot of money along the way. Their efforts had gone well beyond a successful enterprise; they now stood at the helm of a scientific, cultural, business empire with outposts in major American cities and occupied a place of prominence in the national consciousness. The Empire of Skulls had arrived.

Despite all this success—or perhaps because of it—Orson was restless. An entrepreneur through and through, he was always on the lookout for new ventures, and he finally found what he was looking for while planning to build a new home for his family. What he found was actually a shape that, he believed, would change the world, perhaps even more than phrenology.

IN 1842, ORSON purchased a property in Fishkill, New York, his wife's hometown. The location would allow his wife and children to be close to family when he had to work at the Cabinet in New York City or travel the country preaching phrenology. With a majestic view of the Catskill and Fishkill mountains and a short jaunt from the Hudson River, the property was Orson's chance to live a rugged, outdoorsy life—a manly existence his father had modeled for him—and to travel downriver on a steamer when he needed to get into the city.

Orson hoped to build an impressive house on the new property, but he knew that would take time, and his family was growing. As a stopgap measure, he decided to put an addition on the site's existing structure. After hiring a carpenter to take charge of the addition, he hit the road on a lecture tour to earn the money to pay for the new square footage.

For some reason, the carpenter decided to put an entry in the new addition, which was not Orson's original design but which Orson's wife approved. When Orson got home from his speaking tour, he realized the addition with an entryway was too far along to stop. He could deal with that, but he made one request: Given that they had already laid the foundation for the addition, could they make it two stories instead of just one? Another story would add minimal cost to the project but would give the family much more living space.

This confusion with the carpenter got Orson thinking about the way homes were being built in the United States. His addition, like so many homes across the adolescent nation, was haphazard instead of strategic. Most Americans didn't put the proper thought into their buildings, he concluded, and the result was slapdash, ramshackle structures that would crumble in a decade. "Why so little progress in architecture, when there is so much in all other matters?" he wondered. Surely there had to be a better way.[7]

Thus began Orson's visionary undertaking to change the way Americans lived. He began reading extensively on architecture, studying homes across the country on his lecture tours, and philosophizing on the matter. He realized that most homes were square or rectangular, yet nature's forms were "mostly spherical. She has ten thousand globular or cylindrical forms to one square one." What's more, spherical forms have one big advantage over square or rectangular forms—namely, they enclose more space, in proportion, "than any other shape." Take the same length of material, arrange it in a circle, and you will end up with more indoor surface area than if you had arranged it in a square or rectangle. Why shouldn't American homes follow nature's geometry?[8]

But the idea wasn't purely about space maximization. With the spirit of reform coursing through his veins, Orson reckoned he could use nature's geometry to create homes that were comfortable, durable, and within reach of the laboring class. In the same way he championed phrenology as a mental science for the masses, he wanted to make sturdy architecture more democratic. In fact, he believed firmly that homeownership was the best way to lift the lower class into a better living situation, not just for them but for their children and grandchildren as well. "However great your privations, however astringent your poverty, get a home first," he directed the American people, "and the greater your destitution, the more need have you of providing a home—no matter how homely—merely as a means of escaping that poverty."[9]

In truth, Orson was one of many reformers at the time dreaming up solutions to the interrelated problems of class, wealth, living, and flourishing. Many of these reform efforts focused attention on the home, given the increasing significance of private life in a nation whose public sphere was more rambunctious and corrupt than ever. For instance, beginning around 1830, a number of farmers started a movement to reimagine the farmhouse and the homestead as sites of harmony among landowners and laborers and as places of renewal for family virtue. Other reformers turned to communitarian living as the solution to political corruption. Robert Owen and his followers, the "Owenites," created a cooperative settlement in New Harmony, Indiana, that blended education, enlightenment, and equality into a new socialist society. Around the same time, a prominent group of laborers in New York launched a "land for the landless" movement that called on the federal government to distribute plots of land for free to the working class so they could build a proper home and future.[10]

These utopian dreamers and stalwart reformers believed, as Orson did, that a realignment of values around homes, land, and living situations would set the nation on a more just, equitable course. Yet none of these efforts had the kind of long-term influence that Orson's had. He approached the situation with a kind of self-help, do-it-yourself mentality that Americans, wherever they lived, could understand and implement. Unlike many other housing reform movements, which hinged on broad cooperation, governmental intervention, and widespread structural change, Orson's plan simply required that ordinary Americans harness nature's geometry.[11]

The key to it all was the octagon. If you've ever driven around the country, primarily in the Northeast or Midwest, and seen a house shaped like an octagon, you can thank Orson Fowler. To be sure, the octagon had been used in architecture before he came along. Thomas Jefferson had often experimented with octagons in his drawings, including plans for an octagonal chapel. His own home, Monticello, which Orson likely did not know about when he crafted his plan, contained octagonal rooms and angles, but the house itself was not an octagon. Jefferson's retreat home, Poplar Forest, is largely octagonal but also contains rectangular wings that would have offended Orson's sense of harmony. Ultimately, the octagonal home design that exploded in popularity in the 1850s was almost wholly Orson's doing.[12]

He embraced the octagon because it was supposedly a perfect expression of truth, beauty, and utility—the primary architectural principles of the nineteenth century. And even though he had absolutely no training in architecture, even though he had never built a house before, he nonetheless figured he could write a book extolling the wisdom and virtues of his idea. *A Home for All; or, a New, Cheap, Convenient, and Superior Mode of Building*, the first of Orson's two treatises on building octagon homes, was published by Fowlers and

An octagon cottage, which Orson believed practically everyone, especially the poor, could and should build

Wells in 1848, before the eldest sibling had finished his own home. A relatively brief treatise at slightly over ninety pages, the book was predictably engaging and brash, as Orson proudly staked out his place in the pantheon of world-changing architects—his lack of experience be damned.[13]

The book included a lengthy critique of existing home designs—long, narrow ranch houses; rectangular multistory houses; mansion-like houses with "wings"; and quaint, supposedly charming cottages. All of them were problematic, Orson told readers, because they disregarded nature's forms. And he had the math to prove it. Laying out several equations in the pages of the book, he showed how a house with a circumference of 128 feet would result in 1,024 square feet if put into the shape of a square, but 1,218 square feet if laid out as an octagon. That's one-fifth more area for the same amount of house. The gain is even more when compared to a fancy winged house. The octagon, he calculated, will enclose two and a half times more space as a winged house of the same circumference. A winged house might have been good for aristocrats of Europe who wanted to waste their money, but the octagon was best for hardworking Americans who valued their money and living space.

Yet space was only part of the octagon's superiority; it was also cheaper to build. An octagon house with the same overall wall length as a winged house would cost $2,000, Orson estimated, compared to $5,000 for a winged one. "All this two hundred and fifty percent saved, just by the octagonal form, over the winged!" On top of that, the interior layout of the octagon house was the most functional, logical, and democratic possible. "Beauty and utility are as closely united in architecture as they are throughout all nature," Orson wrote. "The octagon form is more beautiful as well as capacious, and more consonant with the predominant or governing form of nature—the spherical"—than all other viable house shapes.[14]

That's what Orson spent the rest of the book proving to readers. The marriage of form and function was evident as soon as visitors stepped inside the octagon, which actually happened in the cellar, not on the main floor. Entering in the lowest level allowed people to hang up jackets, change shoes, and tidy themselves up before heading upstairs. This kept the dirt from outside contained in the basement. Rooms in the lowest level were dedicated to general storage, lumber, wood, milk, and sauce, plus a washing area and a full kitchen. In addition, the cellar featured a central furnace, which burned wood and distributed heat throughout the house. This was far superior, Orson insisted, than fireplaces in each room, as it required less maintenance, was cheaper to operate, and produced better, more consistent heat.

The main floor of the octagon was palatial and efficient. It featured a front parlor, a sitting room, a back parlor, a dining room, and a host of closets. Total square footage of this floor was 1,116. Orson was particularly pleased with the layout of rooms, which was more logical than the layout of rooms in a square home or a winged mansion of similar size. The octagon design saved people steps as they traveled from one room to another, a distinct benefit to women and housekeepers who traversed the area numerous times per day. "Instead of being separated," he wrote, "all the rooms are united, so that you can go from one to another without being obliged to pass through a cold and cheerless entry." It was also great for entertaining, given how close the spaces were. Guests could move from room to room as the evening dictated, instead of being confined to one space.[15]

When it came to entertaining, the kitchen, which Orson called "the stomach of the house," was situated to enhance familial relations. Instead of having to trudge out to a kitchen in the back of the main house, as was common at the time, the woman of the family had only to pop downstairs, grab food and drinks, and return to the gathering. She was thus better integrated into the festivities. "Instead of feeling like she is going away off alone out of doors," Orson wrote, she "feels that she is only a step removed from the rest of the family. What say you, wives, to this?"[16]

The third story of the octagon plan was all about flexibility. Featuring seven large rooms with sizable closets and totaling over 1,200 square feet, the floor was big enough for several bedrooms, a library, and even a guest suite, which allowed visiting parents and children to sleep in separate rooms while sharing a sitting area. This provided guests a space to call their own during their stay. Orson even plotted how light would travel through the floor and facilitate reading in the evenings, without the need for fireplaces or lanterns.

If that weren't enough, there was still the top story of the home, along with a domed roof that provided a proud, prominent facade. The top story was one large open room, 668 square feet, and included storage along the walls. It was perfect for a home gym, a dance studio (which Orson suggested all families should have), and a children's playroom. Dedicated spaces for exercise were indispensable for the health of individuals, families, and even the nation. "How many a debilitated constitution" these activity rooms would revive, he opined. "How many hopeless invalids, now dying by inches, would such rooms in our buildings restore to life, health, and happiness. How many a child saved from a premature grave!"[17]

The most enriching feature of the octagon plan was the piazza, which wrapped around the entire house on every level. Thus each story provided residents the ability to walk around the outside of the home and to find the best view of the morning, evening, and every other time of day. Whenever you needed to feel a breeze or to find shade, the octagon had a space for you. Orson himself gushed about communing with God on the piazzas:

> For one, I dearly love sunrise and sunset. They diffuse through my whole being so sweet and holy a calm, as literally to ravish my soul with earth's sweetest pleasures. 'Tis then I love to retire within my own soul, and draw near to my Marker, imbibe his spirit, bask in the smiles of his love, and open my soul to the influx of divine ideas and feelings.[18]

All told, Orson's octagon house plan resulted in 6,278 square feet. A square house with the same outside wall span would come to only 5,324 square feet. The octagon's extra square footage was in addition to a stronger, better laid out, more efficient, and cheaper design, allowing all Americans, but especially members of the lower class, to secure peace, harmony, and even wealth across generations. "A great public good," he declared, the octagon is the secret to "cheapening and bettering houses" for everyone. You're welcome, America.[19]

TO SOME, ORSON'S obsession with architecture and home improvement seemed disconnected from phrenology, but it wasn't. It was actually a direct extension of his understanding of human nature. On the very back of the head, opposite the eyebrows, sat a cluster of phrenological organs integral to Orson's home-building project. The most important of these organs was Inhabitiveness, which involved love of home, connection to dwelling, even patriotism. Inhabitiveness, Orson wrote, is "adapted to man's need of an abiding place, in which to exercise the family feelings."[20]

Another organ clustered at the back of the head was Adhesiveness, which had to do with social connections. According to the

Fowlers, the best way to develop "social feeling" was to "entertain friends" at home, as opposed to meeting at a bar or attending a dance party. Many public gathering places were nests of depravity, which made entertaining at home a moral as well as a social issue. That was one reason Orson planned for multiple parlors in the octagon house; bringing friends to safe, virtuous spaces was a necessity in the messy modern world.[21]

Still another organ that connected home building directly to phrenology was Philoprogenitiveness, or "parental love." Ensuring that adults were mindful of "children's need of parental care and education," this organ encouraged families to learn, play, and exercise together. Thus Orson designed rooms and layouts that facilitated healthy associations among children and parents, particularly mothers, whose Philoprogenitiveness was, on average, larger than that of fathers.[22]

This phrenological commitment to helping families and to improving home life was not limited to Orson. In fact, several phrenologists at the time took a keen interest in making the domestic sphere a place for proper phrenological development. Foremost among them was Lydia Fowler. Shortly after she married Lorenzo, but before she went to medical school to become the second female doctor in the United States, Lydia grew intensely curious about the implications of phrenology for women and children. After thinking, researching, and writing on the topic for a couple years, she began to publish her findings on childhood development, including in an 1847 article titled, "Phrenology for the Use of Children in Schools and Families." The article was notable not only for the way it connected phrenology directly to the experiences of children but also for the way she described her scholarly mission: "One of the great errors of all authors is that they do not simplify. They pre-suppose that their readers understand more of the various subjects treated of than is actually the case, and hence are not easily followed or understood." Lydia aimed to remedy this defect with the publication of her first book—a two-volume treatise called *Familiar Lessons on Physiology and Phrenology*.[23]

The book was written explicitly for children in the hope that they would grow up with sound phrenological knowledge and, one day, carry the science forward for the next generation. Lydia began with a prefatory appeal to adults that framed the significance of her undertaking:

> Parents and teachers, the minds of children are placed in your hands to mould and direct! They have, as all *must* allow, *natural* tendencies of mind, *natural* propensities, *natural* predispositions; . . . but they *can* be so trained, cultivated, or restrained, that their influence is often greatly modified or entirely counterbalanced. Will you train them for usefulness and happiness or will you suffer the tares of ignorance and vice to grow and expand in their little minds, till they eventually root out all the good?[24]

The rest of the book was designed to captivate young minds. Replete with playful illustrations featuring people and scenes to which children could relate, she took young readers on a tour of physiology and phrenology in a way that made the sciences meaningful to them. The tour was structured, in part, through dialogue between Lydia and imaginary kids. She asked, for instance, "Did you ever hear, children, of Physiology and Phrenology?"

"No," she responded in the voice of little Clara, "*I* never did."

"Clara, do you like to be sick? Do you like to have your head and body filled with pain, and to be obliged to lie on your bed all day long?"

Of course not. And to avoid such a life, all little Clara needed to do was to follow phrenology. Self-improvement was in her hands as much as in the hands of adults.

To make the lessons as appealing as possible, Lydia included vivid examples designed to connect with the target audience. There were examples of kids and toys, parents and siblings, pets and games, school and recess, and similar youth-oriented subjects. In volume one of the book, she offered these examples to illustrate the stomach, digestion, bones, muscles, skulls, brains, teeth, skin, heart, lungs, secretion, fluids, and animals—all grouped under the science of physiology. In volume two, she finally got to the meat of the matter: "It may seem strange to you, children, that any one can tell by the shape of the head whether a man is good, kind, or benevolent; but if you will give me your attention, I will try to make it so clear that you will be able to understand it."

A "guide to self-knowledge" for the nation's future leaders, the volume on phrenology included a section on the founder of the science, Franz Gall, and Gall's trusty assistant, Johann Spurzheim. After the history lesson, Lydia provided a catalogue of the phrenological organs, complete with definitions, locations, explanations, and illustrations. The illustrations were, in fact, essential to the book's appeal. They featured carefully engraved scenes of kids, families, fights, animals, nature, society, and more, all designed to show the applicability of the phrenological organs to young readers. For instance, Lydia explained Amativeness under an illustration of cupid with his bow and arrow. Describing the organ as "feelings of love," she couched the discussion in the way little boys loved their mothers and sisters and the way little girls loved their fathers and brothers. She did mention that when the organ is perverted, it can make "men and women very unhappy." But she simply nodded gingerly to the reproductive function of Amativeness: "When you are older you will understand more about it." The sex stuff could wait.[25]

The illustration for Philoprogenitiveness featured a happy family, standing in their Victorian parlor, the mother looking lovingly into a child's eyes, two cats playing on the ground. Lydia followed the illustration with an anecdote about a man in Schenectady "who was exceedingly fond of pets and children." With very large Philoprogenitiveness, he "frequently went about the city with two little dogs in his overcoat pockets, and one in each hand; and was always surrounded by children."[26]

The discussion of Secretiveness included an engraving of a cat sneaking slyly up to a mouse, ready to pounce on and devour the prey. Lydia followed up the illustration with a story of a little girl faking an illness so she could stay home from school. The joke was on her, however, when "one of her schoolmates came into the room, and told her that the teacher had given them the day for play." Had she been truthful and gone to school, she could have been playing with her friends from the start.[27]

Illustrating the more abstract phrenological organs was something of a challenge, but Lydia found brilliant ways to put abstract ideas into concrete terms. To illustrate Hope, she presented an engraving of a woman holding an anchor and standing on shore near some rocks. She paired the scene with the story of a hypothetical mariner who bravely sailed through a bad storm: "When the winds blow around his ship and the angry waves dash against it with fury, and drive it on the rocks, why is he not filled with despair, when wasted on his frail bark? There is a gleam of Hope in his soul." To illustrate Benevolence, she showed a woman and her three children standing around the bed of a dying old lady, showing compassion as the elder departed this world. To make the point more understandable for readers, she then explained how wealthy children in New York City saved their pennies, "instead of spending them idly for candy, etc.," to buy clothes and school supplies for "many poor and ragged children." Such is true benevolence, she concluded.[28]

Unsurprisingly, the book ended with the kind of moralistic, self-responsibility appeal found in almost all Fowler writing:

> In view of all that you have learned, what do you now intend to do, children? Will you heed all the previous lessons, reflect on them, improve yourselves, take care of your bodies and minds, that you may become useful and influential members of society? . . . Be careful when you decide, to decide correctly—for your own happiness, your present and future welfare, depend on your decision.

Although Lydia was discussing phrenology with children and Orson was designing houses for the masses, their efforts were two sides of the same coin. For practical phrenologists, the home was a place for the cultivation of individual, familial, and civic health. It was at home—at a properly designed and constructed home—that ordinary citizens, young and old, could attune their minds and bodies to truth and live lives of democratic virtue for the sake of an adolescent nation.[29]

JANESVILLE, A SLEEPY Wisconsin village in 1850, was an unlikely locale for architectural ingenuity, but Orson believed it contained exactly what he needed to spread the octagon around the world. In the area to deliver lectures on phrenology, he heard of a building method pioneered by Joseph Goodrich that involved constructing walls out of concrete. Though commonplace today, concrete was a novelty and marvel at the time. By mixing lime, gravel, and sand, builders could erect structures that were supposedly stronger and more durable than logs, wooden boards, stone, and even brick.

When Orson had some free time between Janesville lectures, he made a point to visit one of Goodrich's concrete structures. There to greet him was Goodrich himself, who not only explained the many benefits of his building method but also invited Orson to test the durability of the walls by pounding on them with a hammer.

Bang, bang, bang. Orson pounded away on the cold concrete, feeling the strength of the walls with every reverberating blow.

Did he want more proof? Goodrich offered Orson the chance to bludgeon the walls with a sledgehammer, as hard as he wanted. All he had to do was pay six cents per strike, which would allow Goodrich to repair the cosmetic damage to the inside of the structure.

Boom, boom, boom. The strikes of the sledge seemed to shake the foundations, but the walls themselves were resilient. With every blow, Orson saw a clearer future for his home-building revolution. The walls of octagon homes, starting with his own, needed to be made of concrete.

Returning to upstate New York, Orson got to work building concrete walls and revising and expanding his book around the "gravel wall" method. This method was cheaper and stronger than the board method Orson had promoted before. In the previous edition of the book, he advised people who were building their homes to create walls by stacking wooden boards, anywhere from one to two inches thick, on top of each other, affixing them with nails, and sawing them to the precise measurement needed for a given span. But wood decays over time, he knew, and is prone to fire. Not so with concrete, which would result in durable, resilient homes that could stay in families for generations.

The revised and expanded edition of the octagon house book, published in 1853, shortly after he completed his own home, was titled *A Home for All; or the Gravel Wall and Octagon Mode of Building New, Cheap, Convenient, Superior, and Adapted to Rich and Poor.* In the new edition, Orson explained to readers how the materials they needed for gravel-wall building, particularly stone and lime, were given to humankind by nature and were plentiful across the country. He laid out the right ratios for mixing stone, sand, and lime and how to position mortar beds for most efficient construction. He also explained how to properly frame windows and doors and how to create cheap and effective scaffolding to build the upper levels of the house.[30]

Even with the gravel-wall method, octagon houses were cheap to build. He reported that he was able to build the gravel walls for his home in forty-four days, all for a total of seventy-nine dollars, which was one-tenth of what it would have cost to build brick walls. Better, cheaper, stronger—this was a wall-building revolution for the democratic masses.[31]

The octagon portion of the new edition contained the same comparisons with other house shapes as the previous edition. But this time Orson had examples of actual octagon structures that had been erected because of his advocacy. There were his neighbors in upstate New York who built an octagon barn on their property. There was William Howland, the engraver for Fowlers and Wells publishing, who built a sturdy and efficient octagon cottage. There was a three-story, 6,500 square-foot octagon that an acquaintance of his was in the process of building. And there were floor plans for an octagon

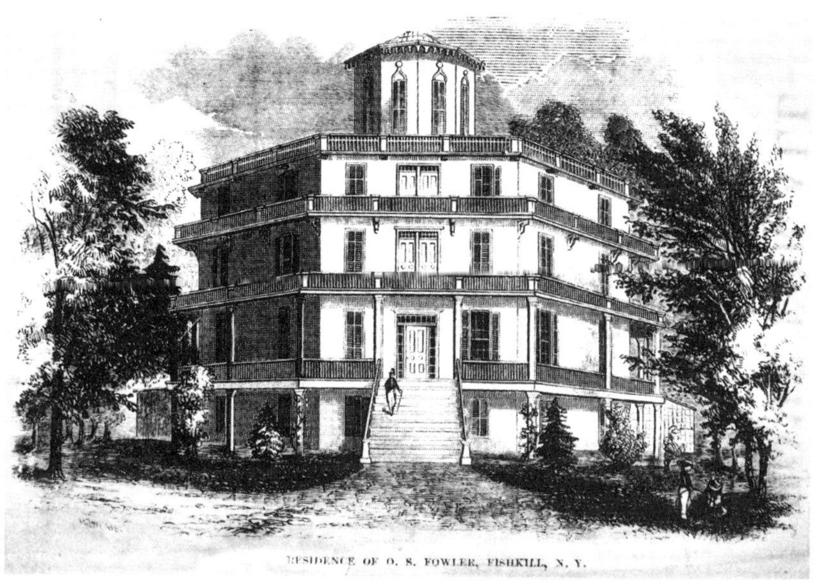

Orson's sixty-room octagon mansion in Fishkill, New York

schoolhouse, an octagon church, and variations of standard octagon homes.

But the most fascinating part of the revised and expanded edition was the tour Orson provided of his own octagon mansion. A magnificent sight to behold, the house featured sixty rooms, and "not one too many." The cellar included, among other spaces, a two-story icehouse; storage rooms for fruits, vegetables, milk, perishables, coal, and wood; a washroom; a kitchen; a furnace; and a dining room for the day laborers who kept Orson's property humming along. The main story of the house included a drawing room, parlor, dining room, and multiple entertaining rooms. There was also a spare bedroom, a sitting room, a library, and a writing room that Orson called his "prophet's chamber."[32]

Above the main floor, the octagon mansion featured two upper stories, plus a domed room that provided amazing vistas of the surrounding landscape—rivers, pastures, hills, and mountains. Each upper story contained eight large rooms, closets included, that could be used for bedrooms, studio rooms, reading rooms, exercise rooms, and playrooms. There were also four smaller rooms, which Orson called "dark bedrooms," as they had no external windows but were useful for visitors or naps. Four separate staircases, including one beautiful spiral staircase in the middle of the house, led from the main story to these upper levels, allowing residents and servants easy transit from bottom to top. Dumbwaiters brought food from lower levels to upper levels, and a carefully planned ventilation system kept fresh air moving throughout the house.

To top it all off, Orson's octagon mansion had its own indoor plumbing system, which carried filtered water throughout the house for drinking, washing, and bathing. On the top of the house were cisterns that ran rainwater through a filtration system designed by Orson himself. Lead pipes then carried water from the cisterns to every relevant room in the house—bedrooms, washrooms, kitchens, even indoor water closets, where residents and guests could heed nature's call without venturing outside. The water and waste from the plumbing system ended up in a receiving box in the cellar, where proper ventilation ensured that the smell would not permeate the rest of the house. Contemplating what ended up in the receiving box, Orson announced to readers that he looked forward to the time when human excrement, "saved from all our cities, villages, and families, would wonderfully enhance the abundance and cheapness of food, and some day it will all be saved."[33]

CONVINCED THAT THE gravel-wall method of octagon building would change the world, Orson kicked the Fowler publicity machine into high gear to promote the revised and expanded edition of the book. *The American Phrenological Journal* printed several lengthy chapters from it, while other periodicals covered it in a way that echoed Orson's boasts of architectural revolution. *The Literary Union* placed his plan for "reform in house-building" ahead of similar efforts and praised his design as "tangible to all and practical *for each individual alone*, and therefore for the masses." Noting the book's "valuable original ideas, which we have no doubt will be of service to many builders," the magazine *Circular* lauded Orson for promoting associative community based on larger homes built closer together, instead of small houses in rural isolation. Another periodical, *The Christian Parlor Book*, called the design "rich in appearance, yet it is a style not beyond the reach of moderate means."[34]

Of course, not every commentator fell in love with the design. *The Country Gentleman* debated how original it actually was, while a correspondent for *The Cultivator* took umbrage with Orson's boasts that his plan was "the best plan" for building a home, which the correspondent called "very undesirable." Perhaps the most biting coverage appeared in *The Anglo-American Magazine* for March 1854, which printed a humorous dialogue among the characters Laird, Major, and Doctor. They debated the merits of the octagon design, which they found rather ridiculous, before proceeding to ridicule Orson as a "bump-hunting land louper" and "one of the most flatulent quacks which this entire age has produced."[35]

Yet all this publicity, positive or negative, invited countless Americans to consider the octagon design for their home, and somewhere between 500 and 1,000 people actually built one. In the wake of Orson's revised and expanded edition, octagon houses started popping up in major cities like New York, Boston, Chicago, Baltimore, and Washington, D.C. They appeared in medium-sized cities like Saint Paul, Minnesota; Madison, Wisconsin; Lewiston, Maine; Augusta, Georgia; Davenport, Iowa; and Salt Lake City, Utah. And they dotted the landscape of smaller towns such as Jeffersonville, Kentucky; Marion, Iowa; Elyria, Ohio; and Middletown, New York. Even the socialist commune of Ceresco, Wisconsin, had an octagon house, as did the progressive reform settlement of Hopedale, Massachusetts.[36]

While many reformers believed Orson when he called this method of building a revolution, one of his associates resolved to prove the point. In 1855, Henry Clubb, a minister, abolitionist, and journalist friend of the Fowlers, had a dream of building an entire octagon town. The shape, he reasoned, would be perfect for creating new communities on the plentiful land out west. With an octagon-shaped settlement, farmers could live close to each other at the front of a wedge portioned out of the overall octagon while their sizable farms could fan away from the houses and still yield rich rewards. With farmers living as neighbors, they could work in common cause and share in the fruits of the new octagonal society.[37]

Determined to make his dream a reality, Clubb saw Kansas territory as the perfect place for the settlement. Kansas needed progressive reformers like Clubb and his ilk, given the ongoing border war between proslavery and antislavery settlers then raging. Relocating a bunch of East Coast abolitionists to the territory would add numbers to the antislavery side of war and, hopefully, lead to a popular vote that brought Kansas into the Union as a free state.

Yet Clubb's community wouldn't be only an antislavery settlement; it would also be thoroughly, proudly vegetarian. A reformer passionate about not eating meat, as the Fowlers were for various stints in their lives, Clubb envisioned the community as a flesh-free utopia that would demonstrate to the world that equality, health, and justice could proceed in lockstep. He thus created the Vegetarian Kansas Emigration Company, a joint stock company that investors, particularly future residents, could buy into to support the new settlement.

With the company formally established, Clubb set about publicizing his plan, and for that he turned to his friends the Fowlers. He wrote a pamphlet describing his vision for the vegetarian octagon community, which Fowlers and Wells printed and sold for twenty-five cents. Originally, he estimated that upward of sixty people would support the venture out West, but because of the Fowler publicity machine, he quickly found himself inundated with interest; around 3,000 contributed to the scheme by buying shares in the Vegetarian Settlement Company.

This was great news, but it brought a new problem: Many of the 3,000 supporters were not die-hard vegetarians like Clubb. They were dedicated reformers, to be sure, but they couldn't give up their chicken and steak dinners. After thinking the matter over, Clubb and a friend, Charles DeWolfe, created a new umbrella organization called the Octagon Settlement Company, which would coordinate numerous octagon settlements out West and allow each one to determine its own principles for living. The company, Clubb wrote, "shall embrace persons of *every reformatory character*, whether Physiological, Vegetarian, Temperance, Hydropathic, or Phrenological, etc.; also, of various religious denominations desirous of locating around their own church or meeting house, in convenient proximity to each other, for which the octagon plan is peculiarly adapted."[38]

By early 1856, it was time to make the octagon settlement a reality. Clubb went to Kansas first to prepare the acreage he had secured, while around one hundred people headed to St. Louis to await Clubb's signal that the settlement was ready for them. That signal came in early April, and the group traveled together three hundred miles to Kansas, where they arrived in May. The plan was to create two settlements right across the Neosho River from each other. One would be for the vegetarians, the other for reformers of various stripes. Hopefully other progressive settlements would follow.

That's when the dream of octagon living came crashing down. When the group arrived, they realized that Clubb's many promises about the area were simply hot air. He had promised that people would arrive to infrastructure, including a sawmill and a grist mill. Neither was even started. He had promised temporary housing for people while they built their own homes. Again, nothing was ready. He had promised a beautiful landscape that would steel their resolve. What they found was squishy terrain full of mosquitoes and rattlesnakes. Some of the settlers left soon after they arrived. Others tried to make it work, but eventually accepted that this utopia, like all utopias, was doomed. In the end, the grand octagon-settlement experiment lasted less than a year.[39]

Octagon dreams were dying—and unfortunately not just on the plains of Kansas. Right as Clubb watched his vegetarian utopia peter out, Orson confronted the beginning of the end of his beloved octagon mansion.

IN 1857, A financial panic—which involves a rapid drop in the value of assets and a rush to withdraw money from banks and investment firms—wreaked havoc across the country, destroying fortunes, decimating families, and debilitating livelihoods. Many building projects, including octagon constructions, were put on hold, never to be restarted. Orson himself was spread thin financially, and when the panic hit, he found himself between a rock and an octagon house. Needing cash but hoping to hold on to his palace, he rented the octagon mansion to a real estate broker for $2,500 per month. The broker turned the mansion into a boardinghouse.[40]

That's when the real disaster started. Even though Orson had heralded gravel walls as the greatest innovation in construction materials for generations, they proved ineffective at keeping out water. Slowly at first, then faster as the house aged, water seeped into the basement and pooled into a bacteria breeding ground that neither the real estate investor nor the residents cleaned up. The cesspool led to an outbreak of typhoid that sent many renters fleeing or begging for mercy at death's door.

Orson tried to hang on to the property through illness and loss, but the situation was untenable. In 1859, he sold the octagon mansion for $150,000. It was a decent return on his investment, and it gave him a nice nest egg for getting through difficult financial times. But it marked the end of his architectural revolution.

Over the next thirty-five years, Orson's octagon mansion traded hands several times and, eventually, fell into disrepair, a blight on the landscape of the magnificent Fishkill area. In 1888, a reporter for the Brooklyn *Standard Union* visited Orson's former palace and found it "deserted and useless." Still impressive from a distance, the octagon was an utter wreck once visitors got close to it. "The storms howl through it, the roof leaks, and the famous spiral four-story stairway with its balconies is rickety," the reporter wrote. Locals even came up with a clever name for it: Fowler's Folly.[41]

The octagon mansion continued to deteriorate for another decade. Then, in 1897, the town of Fishkill had had enough. Determined to use the land for new development, officials sent a team of dynamiters to bring the structure down. They needed dynamite, newspapers reported, because the concrete walls were still as sturdy as when they were built. They might not have kept out the water, but they proved strong and durable—just as Orson had promised. Only several carefully planted sticks of dynamite could destroy them. With the blast, the walls crumbled along with Orson's exciting, strange, overzealous career as an architect for democracy.[42]

CHAPTER NINE

HEADING TOWARD FREEDOM

BEFORE ORSON FOWLER TEAMED UP WITH HENRY WARD Beecher at Amherst College to examine heads, before even Johann Spurzheim sailed for the United States to charm the nation, a minister from Baltimore decided to deliver a series of lectures on phrenology. The year was 1831, and the minister had grand designs on converting the multitudes to this wondrous mental philosophy. To generate publicity for the lectures—admission to which was twelve and a half cents—he printed a handbill explaining that his talks would detail phrenology "according to Dr. Gall, and sundry American Professors," while also covering "the properties and faculties of the mind." The goal was to demonstrate that all people could improve themselves.[1]

The fact that a minister delivered these lectures in 1831 was noteworthy; more remarkable was the fact that the lecturer was African American. His name was Lewis George Wells, and his lectures argued not just that people could improve themselves but, more importantly, that "the mind of a colored person is capable of improvement." More surprising still, he delivered these lectures to "mixed" audiences consisting of men *and* women, Black people *and* white people. Despite the potential for scandal given the mixed-race audience, the lectures were "very successful," reported *The United States Gazette.*[2]

The *Gazette*, in fact, published an extensive account of Wells's lectures, which was then reprinted in several major newspapers, including William Lloyd Garrison's *The Liberator.* Calling him a "colored clergyman of the Methodist Episcopal church (from the south) of high respectability," the paper noted that Wells lectured Monday, Wednesday, and Friday evenings, beginning at 8:15 p.m. every night. After first explaining the science, he moved to "the practical part, taking up all the *powers* and organs of the *mind* individually." It was in this section of the lectures that Wells showed how African Americans could develop their minds to an incredible degree via phrenology. "Those *lectures*," the paper summarized, "are certainly *interesting* and *instructing*; more especially when they emanate from an individual whose opportunities have been so limited; it is hoped the people of this city will avail themselves of his useful instructions."[3]

As *The Gazette* understood, Wells's lectures were significant not just because they explained the benefits of phrenology for Black people but also because Wells himself stood as evidence of the point he was trying to make. He was a Black minister lecturing on a brand-new science that was still in its infancy in the United States. Not only had he mastered the material and its roots in Gall's discoveries, but he understood its practical applicability in America as well, specifically for the enslaved population, and could explain it eloquently and convincingly to a general, mixed audience. The performance itself, then, was an argument: When given their freedom, African Americans could become like Wells, preaching the gospel, mastering new sciences, and captivating the masses.

To be sure, Wells was not alone in this enthusiastic embrace of phrenology as a science of progress for Black Americans. Far from it. Many Americans at the time, both Black and white, heralded phrenology not as a pseudoscience of oppression but as a scientific beacon lighting the pathway to racial equality. This is not to say that there were not still major problems with phrenology's handling of race. But for numerous reformers, it was the best the scientific world had to offer in aid of liberation. Practical phrenology became the conceptual, empirical underpinning of their advocacy—especially when it came to destroying the evil system of slavery.

AS WAS THE case with most ideas in antebellum America, phrenology had a conservative pole and a progressive pole. On the conservative end of the spectrum was the work of Charles Caldwell, one of the earliest American proponents of phrenology, an unapologetic racist, and a proud slaveholder. Living and working in Kentucky in the decades before the Civil War, he used phrenology to argue that white people were superior to all other races because of their "*native* endowments, *intellectual* and *corporeal*." As he explained to his medical school students in 1827, years before the Fowlers had even heard of the science, "Phrenology discloses, clearly and satisfactorily, the cause of the intellectual differences between the several races of men. It shows the ground of the decided superiority of the Caucasian race; and the reason why, when nations of this description have come into collision with those of the other races, they have always vanquished them." On top of that, Caldwell insisted that "the Caucasian race alone" was capable of "real human greatness, marked by the omnipotency, and decorated in all the elegance, of genius." The implication for African Americans was pretty clear to Caldwell. As he explained in a letter to the Scottish phrenologist George Combe, "Depend upon it my good friend, the Africans must have a master."[4]

It's hard to imagine a clearer-cut statement of white supremacy. In Caldwell's hands phrenology authorized white subjugation of other races and conquest of the world. The science also provided, according to Caldwell, stark evidence of polygenesis—the belief that human races have separate origins. Black people and white people, and those from all the other races besides, began in different places and are distinct by nature. In addition, the races are more or less static given their varied origins. Neither the African race nor any other race could ever catch up to Caucasians, whose superior endowments made them natural rulers.[5]

The Fowlers knew and respected Caldwell. He was a champion of their beloved science and an ally to their cause. And in some instances, they made statements that sounded a lot like those of the slaveholding Kentuckian. In *The Illustrated Self-Instructor*, for instance, a pamphlet that was often distributed when sitters paid for an examination, the Fowlers insisted that "the Caucasian race is superior in reasoning power and moral elevation to all the other races, and, accordingly, have higher and bolder foreheads, and more elevated and elongated top heads." Of course, that was just not true—a sweeping generalization about groups of human beings that remains a very real part of phrenology's problematic legacy.[6]

If such stereotypes were the extent of the Fowlers' discussion of race, they would align squarely with the conservative, racist end of the phrenological spectrum. Yet, as with all things phrenology, the story is more complicated. The Fowlers also made many statements that undercut exactly what Caldwell argued. They stood resolutely opposed to slavery and even funded abolitionist initiatives. These seeming contradictions were common at the time and certainly not limited to the Fowlers. But in their worldview, problematic statements about race *and* the fight for liberty and equality could exist side by side because of phrenology's central promise—namely, improvement. Real progress in individuals and groups was always possible, they argued, so even if white people were supposedly superior to other races in terms of reasoning power and moral elevation, it didn't have to stay that way. The possibility of intellectual advancement was baked into human nature, which meant that people had an inherent right to develop their minds properly. Practical phrenology could actually become a strong, useful argument for racial progress.

That's what Orson tried to communicate in his 1843 book on human lineage and development. Titled *Hereditary Descent: Its Laws and Facts, Illustrated and Applied to the Improvement of Mankind*, the book explored the role of ancestry in the progress of individuals, families, and nations. He hoped that people who were thinking about becoming parents would read his work, visit a phrenologist, and think hard about how their characters would combine in their offspring. It was a tough but essential topic for Americans to contemplate, and Orson had no patience for "those who are so very *extra* delicate and refined that they cannot investigate this subject without a blush." Anyone that delicate, he insisted, is too modest to get married in the first place: "If true modesty need not be offended by marriage, it certainly need not blush to learn the duties and relations necessarily connected with, and growing out of, that marriage."[7]

Much of the book concerned how traits were handed down from one generation to the next, both in humans and in other animals. Along the way Orson offered a wealth of examples from his travels as a practical phrenologist, including about families with similarly large heads despite their small stature; kin with six fingers on one hand; siblings with hair discolorations in the same spot; and generations of "porcupine men," whose skin contained distinct bunches of "hairy substances growing out on them quite analogous to the quills of porcupines." Beyond these familial appearances, he explained the way diseases, both of body and mind, were handed down from one generation to the next. He even explained that insanity was as much a disease as anything else, not some divine punishment.[8]

Yet practical phrenologist that he was, Orson insisted that inheritance was not the only controlling factor in one's existence. Education played an important role, and people could always improve on what their parents had given them. As with so many texts in the Fowler canon, *Hereditary Descent* was a book about progress on small and large scales. "The present is emphatically an age of *reform*," Orson wrote.

> The ice of the dark ages, which has bound the river of society and fettered its current since the creation of Adam, is beginning to break up. Mankind are freeing themselves from the shackles of ages, and attempting various reforms in government, politics, the arts, sciences, religion, morals, temperance, etc., and with partial success, but none of the reforms now in progress can extend far or effect much, till they begin with the *root* of vice, and make it a root of virtue—till they commence with the *germ*.

Turning the root of vice into a root of virtue meant understanding what parents passed down to their kids, and the ways that

Orson Fowler in his later years

education could redirect natural instincts. Indeed, Orson reminded his readers time and again, parents have a responsibility to future generations. When choosing a mate, people must consider what their offspring will inherit rather than thinking only of satisfying their "animal indulgence."[9]

Orson's point about reproduction based on a concern for future generations trod on dangerous ground. The logic of the argument was the same logic that appeared in social Darwinism later in the nineteenth century and in eugenics in the twentieth century. Yet Orson's thinking went in a very different direction than these later sciences. After spending dozens of pages discussing different races, including the "colored race," the American Indian race, the Jews (which Orson classified as a nation instead of a race), the Chinese (again, a nation), and the Caucasian race, Orson made a pivotal point: Ultimately, the human race is defined by similarity, not difference—unity, not division. "Slight changes," he wrote at a time when such sentiments were dangerous, "induced by climate and circumstances, appear in different races and ages, but at heart, all appear to have been the same . . . The *oneness* of our race is most apparent. The avenues to the human heart are the same in all." In fact, unity is so controlling in humanity that "he who has learned human nature once, need not learn it again."[10]

This did not mean that there were no differences among the races. Orson offered more sweeping generalizations about head shapes and facial features of different groups. Yet as the book reached its culmination, Orson complicated the matter by confronting perhaps the most taboo subject of the day: race mixing. In fact, just as readers might have expected Orson to make a case for keeping races and bloodlines pure, as many thinkers at the time were wont to do, he argued just the opposite. When people of different races have children, the characteristics of those races create new combinations and modifications for the benefit of future generations, resulting in "new phases of character" and

> physical propensities hitherto unknown, which, instead of dying with those individuals or generations in which they originated, will not only live and spread throughout the countless millions of their descendants, but also form new bases or causes, the product of which will be phases of character and kinds of talent now unknown and inconceivable to mankind.

For Orson, intermarrying was not only permissible but highly advantageous to the progress of humanity, given that "the diversity existing among mankind touching *mental* qualities" was "infinitely greater than that appearing to their looks and other merely *physical* conditions."[11] In other words, whatever changes happened physically when races mixed were of little concern in light of the benefits that humanity received from new mental combinations.[12]

If the point weren't clear enough, Orson turned directly to the kind of race mixing happening in the United States:

> The first child produced by the union of a Caucasian and an African parent was a mulatto, differing in color and form of body, and in cast of mind and tone of feeling, from all other members of the human family. Nobody like him, either mentally or physically, had ever before existed. His children then intermarried, perhaps with whites, perhaps with blacks, and produced children unlike either parent or ancestor, because compounds of two parents the like of one of which had never before existed, and therefore the compound of this unique parent with one unlike himself, necessarily proceed another *sui generis*; and their intermarriages, others possessing a mixture of qualities never before exactly equaled, or if equaled, the conditions and circumstances of the parents and all the ancestors of these two, were not exactly alike . . . And this principle applies to every member of the human family, past, present, and prospective; and hence, mainly, the *diversity* of the human character and physiology.[13]

This was a view of the human family radically different from the view of Charles Caldwell. Although both perspectives were premised on phrenology, Orson seized on the idea of improvement to argue that humanity was stronger through diversity. When Black people and white people had children, they added to the range of bodies and minds that existed, and all of us benefited. The argument was practically unheard of at the time, particularly in antebellum America. Yet Orson's mission to "improve the stock of mankind" meant laying "the axe of reform to the root of this tree of vice and misery" and planting in its place "a root of virtue." Diversity was that virtue.[14]

A SLIGHT, BESPECTACLED man with strong eyes and a high, arching forehead, the prominence of which was accentuated by a vanishing hairline, William Lloyd Garrison was abolitionism's righteous prophet. His advocacy effectively kicked off a movement in the United States that sought both the immediate end to slavery and equal rights for African Americans.

As a prominent reformer interested in remaking America, Garrison had been aware of phrenology for years and first had his head felt in 1836. As the story goes, he did not identify himself when he sat for Lorenzo and Orson in New York. Yet, as *The Liberator*—Garrison's fire-spitting newspaper published out of Boston—noted in publishing his phrenological description: "We see not how the striking accuracy with which the prominent traits of his character are delineated, can be accounted for in any other way, than by supposing the science of Phrenology to be founded in truth." Sliding his fingers across the bald curvatures of Garrison's cranium, Lorenzo found a remarkable head. With "an active mind, quick perception, strong investigating powers, great imagination, great determination and pride of character," Garrison operated with tremendous courage, using "a moral weapon" in his cause instead of "a physical one." And yet his Destructiveness was large, meaning that he was positioned to tear down whatever stood in the way of his moral war. Fortunately his large Destructiveness was properly directed through his fully developed religious faculties. "On the subject of religion," Lorenzo found, "he takes general and liberal views, and is not guided at all by creeds and ceremonies . . . He was born to take the lead, rather than be led. He always engages with his whole soul in anything he undertakes."[15]

Standing in the room at the time to help with the examination, Orson quipped, "You would make a roaring abolitionist."[16]

To Garrison, this phrenological reading robustly confirmed the truth of the new science. And it wasn't just the character description that stood out to him, accurate though it was. It was also the way phrenology muscled through the slings and arrows of its critics.

Garrison saw a striking parallel between phrenology's rise and the abolitionist fight he was leading. "Thus far," he wrote in 1840, "phrenology, like a rough diamond, has grown brighter by attrition. Placed in a fiery ordeal, it has lost nothing by the dross of ignorance and obscurity, in which it has lain hidden for ages." Indeed, the Fowlers' "zeal, enthusiasm, and hope" places them in the company of leading reformers doing their damnedest to change the world.[17]

That was a big reason why Garrison covered phrenology regularly in the pages of *The Liberator*. Over the years he praised *The American Phrenological Journal* for its contributions to human knowledge and encouraged his readers to subscribe to it, along with a host of other Fowlers and Wells publications. He stood up for Orson when David Meredith Reese deemed phrenology one of the greatest humbugs of the age, which was not surprising to Garrison, given that the foolhardy, incorrigible Reese had also lambasted abolition. He reprinted a lengthy account of the time that Orson, while blindfolded, examined the heads of other abolitionists at the American Anti-Slavery Society's office, which was just a couple doors down from the Phrenological Cabinet. He praised the time when the Fowlers attended the Boston Female Anti-Slavery Society meeting and debated a critic of phrenology before publicly examining heads to the delight of the audience. He lauded the phrenological examination the Fowlers gave to the abolitionist and reformer Lydia Maria Child as "very remarkable" and evidence of "the truth of phrenology as an accurate and valuable science." On top of all that, Garrison covered various Fowler lectures and highlighted the ways their science bolstered other reform movements, including temperance, women's rights, and vegetarianism.[18]

Throughout the 1840s Garrison interacted regularly with the Fowlers at various reform gatherings, and they developed both a friendship and a trust, which became particularly valuable to the cause of abolition in 1851. That year Lorenzo agreed to serve as an expert witness in a legal case involving a fugitive slave. John Bolding was born into slavery in Columbia, South Carolina, in 1824, but he managed to escape to Poughkeepsie, New York, as an adult. In 1851, he was working as a tailor and married to a free woman when Robert Anderson, slaveholder from South Carolina, paid residents in Poughkeepsie and federal marshals to get his "property" back. After outrage from locals, the incident moved to federal court in Lower Manhattan, where abolitionists came up with a rather brilliant argument: Bolding was not actually African American. The claim, which was only possible because of Bolding's light skin, hinged on evidence of his lineage.

On August 28, 1851, Lorenzo took the stand in Manhattan to explain. Recounting the time he examined the man blindfolded, without knowing anything about him, Lorenzo concluded that there was nothing in the examination to suggest that Bolding had "any African blood in him." He then held up the skulls of an American Indian and an African in court, demonstrating their differences and noting that Bolding's head was much more like the American Indian skull than the African skull. Despite this testimony, and corroborating testimony from Samuel Wells, the judge ruled that Bolding was the property of Anderson.[19]

Garrison, Lorenzo, and the other abolitionists involved in the case knew that Bolding was an escaped slave. But in response to the Fugitive Slave Law, they also knew they needed to use whatever weapons were at their disposal to keep people free. Lorenzo's testimony used phrenology for advancing a novel argument that Bolding's head was not the stereotypical head of an "African." The argument didn't stick, but it was a valiant effort to use the stereotypes in which phrenology trafficked for a man's freedom.

When the Union claimed victory in the Civil War and began the slow, uneven, difficult process of emancipating slaves, Garrison began the process of shutting down his famous abolitionist newspaper, *The Liberator*. Lorenzo, a longtime subscriber, heard the news and sent a letter of gratitude to Garrison, which appeared in the paper in August 1865. Titled "Letter from L. N. Fowler, the Eminent Phrenologist" and addressed to "My Old Friend, Wm. Lloyd Garrison," Lorenzo's note spoke of the pain he felt upon hearing that *The Liberator* would cease publication, but also the pleasure of the news because the cause for which the paper was founded "no longer exists." He went on to praise Garrison for winning "a martyr's fame by a martyr's labors, bearing opposition, imprisonment and calumny when the cause was weak." Lorenzo even went on to suggest that the paper not cease publication but change its name to *The Liberated* and provide an outlet for former slaves to "tell their experience, give their views, and make known their rejoicings as they become educated."[20]

The friendship and common cause between the great abolitionist and the great phrenologist surfaced again in 1867, when both Garrison and Lorenzo were in England. Garrison had reached out to Lorenzo in the hope that the two men could visit together, but the timing just wasn't right. "Friend Garrison," Lorenzo wrote in reply, "I am sorry I shall not be able to see you while in Manchester . . . You have many friends there. I am thankful with thousands of others that your life has been spared to see the end of slavery in America." Lorenzo then thanked his friend for being "instrumental in the hands of God to bring freedom to the lands and bodies of millions of human beings." He signed off simply, "Your old friend, L. N. Fowler."[21]

OTHER ABOLITIONISTS WHO embraced phrenology included Lucretia Mott, Lydia Maria Child, Elizabeth Cady Stanton, and Susan B. Anthony. Like Garrison, they saw the science's promise of equality through improvement as a foundation for abolition and suffrage. In addition, both Sarah and Angelina Grimké had their heads read and championed phrenology, as did newspaper editor and minister Joshua Leavitt and education reformer and transcendentalist Bronson Alcott, along with many others. Then there was Theodore Weld, who not only embraced the science but enlisted the Fowlers in his abolitionist crusade.

Though Weld tended to shun the spotlight and work behind the scenes on emancipation—unlike Garrison, who loved the spotlight—his contributions to the cause were unparalleled. In the

Lorenzo Fowler in his later years

mid-1830s, he worked for years to collect firsthand accounts of the evils of slavery and finally published a gut-wrenching exposé, *American Slavery as It Is*, in 1839. It was the most influential nonfiction book of the abolitionist movement, rivaled only by Frederick Douglass's autobiography, and it served as primary source material for the most influential fiction book of the struggle. When writing *Uncle Tom's Cabin*, Harriet Beecher Stowe drew extensively upon *American Slavery as It Is* to depict the horrors of slavery.

In the midst of working on *American Slavery as It Is*, Weld got an idea that many Americans at the time got: He should put phrenology to the test. In 1837, as Weld's soon-to-be wife, Angelina Grimké, explained in letter to a friend, Weld visited one of the Fowlers (she didn't specify which one) and "disguised himself as an omnibus driver." It didn't work. "The phrenologist was so struck with the supposed fact that an omnibus driver should have such an extraordinary head, that he preserved an account of it, and did not know until some time after that it was Weld's." As Angelina further explained, this was actually the second time Weld had sat for a phrenological reading. The first one, which happened in Utica, indicated that Weld was deficient in Color, apparently because of a dip in the ridge around his eyebrow. That sparked a childhood memory in Weld of when his mother told him, "Theodore, it is of no use to send you to match a skein of silk, for you never bring the right color."[22]

A couple years later, Weld was working at the Anti-Slavery Society's office, located at 143 Nassau Street, only a stone's throw from the Phrenological Cabinet, and he realized the Fowlers could help him make a new argument for abolition. Flanked by two associates, Weld strode to the Cabinet with a special bust tucked under his arm. Asking Lorenzo to turn away, he placed the bust on a table in the middle of the examination room and covered it with a white sheet. He then asked the younger Fowler brother to examine the object through the sheet and to pronounce its character.

Lorenzo didn't know it at the time, but under the sheet was the bust of a man whose life story would become a powerful argument for the immediate abolition of slavery. Eustache Belin was born in 1773 into slavery in Saint-Domingue, a French colony in what is now Haiti. Forced to work the dangerous sugarcane production facilities of wealthy plantation owner Paul Belin de Villeneuve, Eustache survived long enough to see the Haitian Revolution of 1791. Bloody and successful, the revolution saw slaves rise up against their captors and wage war for over a decade, eventually securing control of the French part of the island and leaving some 300,000 bodies littered across the landscape. Witnessing the violent uprising, Eustache made a choice not to turn on his captor but instead to help Villeneuve and over one hundred others escape the island, demonstrating incredible courage and compassion in the process.[23]

Able to sneak to the port, Eustache arranged for himself, Villeneuve, and dozens of others to hide in a ship bound for Baltimore. Everything went smoothly at first. But once on the high seas, a British cruiser intercepted the ship and discovered the human cargo.

In detention on the vessel with his compatriots, Eustache hatched a plan. Because he was skilled in the kitchen, British officers gave him free rein of the ship and asked him to cook his spectacular meals—rich in flavor, infused with unfamiliar spices. This gave him the chance to pass messages to his detained comrades.

One evening, as the officers were sitting down to dinner, Eustache and a small group of captives, now armed with rusty swords, stormed the dining room. Caught completely off guard, the British officers had no time to get to their weapons and were taken into custody. After deciding to spare the lives of the officials, Eustache and his men sailed the ship safely to Baltimore. Almost as soon as they landed, tales of his bravery reverberated across the world, especially in Europe. Decades later, Eustache was honored in Paris and celebrated as a hero. When Villeneuve eventually passed away, the former slave inherited his fortune.

In 1835, when the famous Eustache was in his sixties, he allowed a prominent French phrenologist to make a cast of his head. The cast was then used to make a bust, which was copied numerous times and placed in a number of high-profile phrenology collections, including that of George Combe. When Combe visited the United States in 1838, he brought Eustache's bust with him, and a copy of it ended up in the hands of Theodore Weld and the Anti-Slavery Society office in New York. Knowing enough about phrenology to realize Eustache's head was special, he hatched a plan to fool Lorenzo and, in turn, to advance the cause of abolition.[24]

When Lorenzo felt the bust under the sheet, he realized immediately that his fingers traced the cranium of an amazing man. So amazing was the head, in fact, that Lorenzo assumed it belonged to a white man. Large in Benevolence, the man's head distinguished "him for good nature and humanity." It also exhibited "principle, moral sense, and strong conscientious feelings." The owner of the head, Lorenzo continued, was "very domestic, affectionate and kind, strongly attached to children, and to home and place." Practical and able to "carry into execution his plans," the man "had uncommon forethought and search of intellect; was a great reasoner—great planner; full of designs; not deficient in powers of intrigue; can plot and plan; uncommon share of ingenuity."[25]

Delighted with the reading, Weld pulled back the sheet to reveal the head of a Black man. It was Eustache, he told Lorenzo, and the famous phrenologist knew immediately what that meant. He had felt the head of a former slave, whose bravery, courage, compassion, and selfless leadership saved the lives of Villeneuve and over one hundred others. With large Benevolence and the ability to reason, plot, and plan, he was able not only to smuggle his compatriots out of Haiti amid a bloody war but to outsmart the British officers and take control of the ship as well, landing it safely in Baltimore. If the head of a slave could develop to such a point that it fooled a famous phrenologist into thinking it belonged to a white man, the future of improvement for African Americans was bright indeed—at least once they were freed from their shackles.

This message of hope and possibility followed Eustache's bust for decades. His head showed the extent to which Black people could develop their minds if given the chance, becoming like the heads of white people. Armed with Lorenzo's reading, abolitionists declared that immediate emancipation and equal rights were the only proper response to slavery. Anything else was an affront to God and human nature.

PERHAPS THEODORE WELD and other white abolitionists put too much faith in phrenology to advance their cause. Indeed, they *certainly* put too much faith in it; the science was not an accurate depiction of the brain and mind, and slavery maintained its stranglehold on the nation for decades more. Yet they were not wrong to see the science as a beacon of equality and a threat to bondage below the Mason–Dixon line. Many Black Americans, including some radical Black abolitionists, saw the same thing.

One of the strongest contingents of Black Americans supporting phrenology came from physicians. Though Black physicians were relatively rare at the time, those who were able to receive medical training often distinguished themselves as community leaders and proponents of phrenology. James Joshua Gould Bias was a preacher, key figure in the Underground Railroad, and medical doctor. After emancipation, he moved to Philadelphia, studied medicine (including phrenology) at the Eclectic Medical College, and obtained his degree in 1852. As a physician, he served the local Black community in Philadelphia and helped many fugitive slaves who found their way to the city, earning a reputation for uncharacteristic kindness and compassion. In addition to treating health maladies, he was also a practical phrenologist who relied on the resources of Fowlers and Wells to conduct his work. For instance, when giving exams he used Orson's "Synopsis of Phrenology" as a character record and keepsake for his sitters.[26]

Another prominent Black physician and phrenologist was Henry Lewis, an American educated in Liverpool who settled in Michigan and worked across the Midwest. In 1846, a correspondent named Nathan Falor sent Orson a note about the incredible effectiveness of Lewis's advocacy: "A negro is now lecturing in those parts who makes converts by the score wherever he goes. In a small town near here, he has been employed about a week, and they will still employ him some time to come. He will act as agent for the Journal if requested."[27]

A couple years later, Lewis embarked on a phrenological lecture tour, speaking in several cities, including Rochester, which caught the attention of a reporter for Frederick Douglass's *The North Star*. Lewis's lectures were well timed, the reporter noted, given that the local Black community was forming a literary society, and "his mental feasts created an impetus" for their work. At the conclusion of his talks, the audience passed a resolution praising "his lectures on phrenology and mesmerism," cheerfully commending "his efforts to the favorable attention of all votaries of science and reform."[28]

Then there was H. Jerome Brown. Born in Baltimore and trained in Philadelphia, Brown took quickly to phrenology, physiology, physiognomy, and related sciences. For much of his career he moved back and forth between Maryland, Delaware, and New Jersey, but wherever he resided, he lectured on and practiced phrenology. Praising his support of the science, *The Christian Recorder*, a prominent African American newspaper at the time, deemed him "one of the most useful men in our ranks," with an "easy and rapid flow of language" and "quite forcible and energetic in his descriptions." In another article, the same paper reported that Brown lectured "to the people of Newport, both colored and white, on 'Phrenology and Physiology,'" and that "all those who were present will long remember this lecture." Perhaps Brown's biggest compliment came in 1862 when *The Christian Recorder* concluded that he "lectures with as much ease, if not more so, than any one we have ever heard upon the subject. He is a second Fowler, though a colored man."[29]

Although phrenology was widely embraced by Black physicians and activists, there were notable detractors, the most prominent being James McCune Smith. Educated in Glasgow and the first Black American to obtain a medical degree, Smith practiced medicine in New York and published a number of influential scientific articles. He also worked alongside Frederick Douglass and other leading abolitionists to create the infrastructure needed to combat slavery.

As early as 1837, the year he received his medical degree, Smith was back in the United States delivering lectures against phrenology. Covering the lectures, *The Colored American*, a paper largely opposed to the science, lauded Smith's lecture in New York for demonstrating that "there are no principles in philosophy, reason, nor revelation upon which they can found their theory." Both Smith and the paper were concerned about the influence of phrenology on the "unthinking community," which would treat "bumps" as their moral destiny. Indeed, Smith's main objection to the science was that it stood opposed to Christianity, cultivating in its supporters "infidelity and irreligion," which he said he witnessed firsthand among his fellow medical students in Glasgow.[30]

Smith's vehement opposition to phrenology was significant given his medical training and leadership. But his conception of phrenology grew largely out of the way it had developed in Europe. Across the Atlantic, phrenology did tend toward "irreligion" and viewed a person's mental makeup as more or less static. But in America the Fowlers' practical phrenology meant something much different. In their hands the science conformed with God's revelations and extended the hope of progress to everyone. Black physicians who learned the Fowler version of phrenology had a much different take on it than Smith.

Perhaps the best example of this was the physician, journalist, and abolitionist Martin Delany. Recognized today as the "father of Black Nationalism"—a movement later championed by the likes of Marcus Garvey and Malcolm X that promoted Black pride, self-determination, and economic self-sufficiency—Delany was an astute political actor, promoting voluntary emigration to Africa so that Black Americans could start their own nation-state apart from the racist structures of the United States. His 1852 book, *The Condition, Elevation, Emigration, and Destiny of the Colored People of the United States*, was a clarion call for racial solidarity in the wake of the Fugitive Slave Act, which mandated that Northern states assist Southern slaveholders in getting back their escaped "property."[31]

Before promoting Black Nationalism, Delany wrote about and practiced phrenology. Cofounder with Frederick Douglass of *The North Star*, a paper written, edited, and published by Black Americans to tell their stories, Delany traveled the country to document initiatives and communities largely overlooked by the mainstream press. On one such sojourn in 1848, he stopped in the town of Troy, Ohio, and met a remarkable fourteen-year-old boy named Simon Foreman Laundrey. Laundrey was Black, but more importantly, and despite his age, he was a devotee of the new science. Delany called him "a natural Phrenologist, who examines heads, reads out the organs, and delivers lectures on the science. He has had but comparatively little schooling, and what his qualifications are I do not know." Yet Delany could see that young Laundrey was a true genius whose "examinations compare well with experienced and competent professors of the science of phrenology." Impressed with

Martin Delany, pioneer of Black Nationalism and a committed phrenologist

the teenager's work, Delany sat for a reading, which the young man conducted with aplomb, passing "his little hands over the organs" and "reading them with as much facility as Fowler."[32]

Delany felt comfortable praising Laundrey's phrenological prowess given his own medical background. In 1833, Delany began his medical training as a physician's assistant in Pittsburgh during a cholera epidemic that ravaged the city. Over the next decade, he continued to study medicine while also launching a newspaper, *The Mystery*, which survived for only a few years. Then, in 1850, Delany got his big break as one of three Black men admitted to Harvard Medical School. However, after only a month, white students at the institution started protesting the presence of Black men among their ranks. After some consternation, Harvard officials caved and kicked Delany and the other Black students out of the school.

After Harvard, Delany returned to Pittsburgh, fully aware that the white power structures of the day would not let him advance in the ways he knew he could. His eyes turned toward Africa and his mind toward Black Nationalism. All the while, he continued to view phrenology as a science of possibility, not oppression, praising it in his newspaper columns and writings and affirming the phrenological undertakings of Henry Lewis, J. J. Gould Bias, and others. For the father of Black Nationalism, the science of bumps, spans, and skulls was radical, progressive, and hopeful at a time when Black Americans needed a lifeline.

WHILE DELANY EMBRACED phrenology as a science of progress and James McCune Smith decried it as wrong and dangerous, Frederick Douglass, the most influential Black abolitionist of the nineteenth century—and one of the most influential Americans of any century—fell somewhere in between. In fact, he was closer to Delany's support than to Smith's rejection. He just knew that phrenology was more complicated than it first appeared.

Born into slavery in Maryland in 1818, Douglass learned to read and write during his teenage years. He was particularly fond of a book called *The Columbian Orator*, an anthology of speeches and essays from the late eighteenth century designed to model for students proper, effective discourse. The anthology showed Douglass how to make arguments for freedom, how to structure his ideas, and how to adapt his appeals to particular audiences. Little wonder he developed a prominent brow ridge, which indicated penetrating perceptions of the world; a notable crest under the eyes, which indicated a penchant for powerful language; and a high, robust forehead, which indicated a well-developed sense of beauty, human nature, and morality. With such a head, it's unsurprising that he escaped from slavery and headed north.[33]

After his perilous journey to freedom in 1838, Douglass ended up in Massachusetts, where he fell in with a group of abolitionists and continued to develop his intellect. He subscribed to William Lloyd Garrison's *The Liberator*, which set his mind ablaze with its fiery, unflinching attack on slavery and its denunciations of evil in both the South and the North. After meeting in person at an antislavery meeting, the two men saw tremendous promise in each other. Garrison perceived someone whose personal story was so eloquent and compelling it was sure to propel the movement; Douglass perceived someone whose religious devotion to ending the evil system would change the nation. Becoming fast friends and collaborators, they preached abolition with the divine calling of prophets who warned a fallen people of God's impending judgment.

In part because of his early relationship with Garrison, Douglass approached phrenology with an open mind. He read Garrison's coverage in *The Liberator* and saw the way other abolitionists, including his friend Delany, used it. *The North Star* covered the science regularly. In 1848, for instance, Douglass printed Orson Fowler's prospectus for *The American Phrenological Journal*, which had been around for a decade by that point. Yet in *The North Star*, the prospectus operated in a new way, inviting Black readers to investigate a science of equality. "To reform and perfect ourselves and our race, is the most exalted of all works. Yet, to do this we must understand the Human Constitution. This, Phrenology, Physiology and Vital Magnetism embrace, and hence expound all the laws of our being, conditions of happiness, and constitute the philosopher's stone of Universal Truth." A science of "our race," the human race, phrenology was for everyone eager to know truth and to advance themselves.[34]

Douglass offered his own take on phrenology, albeit obliquely, in an 1851 article titled, "Cuba and the United States." Warning readers of a surging American imperialism, he decried "our voracious eagle . . . whetting his talons for the capture of Cuba." The government's drive to conquer Cuba, he believed, was yet another land grab akin to the acquisition of Texas, New Mexico, California, and Florida. In these conquests politicians schemed with slaveholders and

Frederick Douglass, the most photographed American of the nineteenth century and a keen assessor of phrenology

masked their bloodlust and greed with shouts of "Liberty," which Douglass called the "watchword and disguise of freebooters, pirates, and plunderers." He then offered a telling analogy—one about the development of the national mind: "It is laid down as a truth in phrenology that an organ of the mind increases, both in size and activity, by exercise; and we are told by moralists, that the first step in the broad road of transgression, is taken with more reluctance than any on this side of the last dread leap into ruin. If this be so, a career of high handed villainy is opening to our Republic."[35]

The analogy between U.S. foreign policy and phrenology was straightforward enough but also revealed Douglass's complex views on the science. His statement about exercising an organ of the mind and increasing its power was neither an endorsement of phrenology nor a dismissal. He simply conveyed a "truth in phrenology" without saying it was, in fact, true. Yet he also used the science to advance his argument, depicting a national mind with an organ of conquest. The more America acquired new land, the stronger its hunger for expansion became. "If this be so," he said—still an open question—the United States would surely continue its piratical ways.

This openness to phrenology remained a part of Douglass's work for several more years. In 1852, he published a serialized novel by J. R. Johnson called *Uncle William's Pulpit.* Part of the novel included an exchange between Uncle Williams and his niece Julia, who makes an impassioned defense of phrenology. In response to Uncle Williams's argument that phrenology "leads to infidelity and fatality," which was also Smith's argument, Julia insists that it concerns "one of God's most glorious works for its text, I mean MAN." After another warning from Uncle Williams about the science's dangers, Julia retorts: "I thank you much for your warnings; but I see no danger of being held into infidelity, while all my investigation of phrenology expands my views, elevates my affections, and causes me to feel every moment that in God I live, and move, and have my being." The narrator of the story follows up the exchange by noting that Julia must "have made herself conversant with Fowler's works."[36]

A year later, in 1853, Douglass published a notice of upcoming lectures by Orson Fowler in Rochester's Corinthian Hall. Living and working in Rochester, Douglass himself may or may not have attended the lectures, but he did tell his readers about the topics Orson would cover and noted that at the conclusion of the series, "All examinations made blindfold if desired." Whatever Douglass personally thought of the science, he thought enough of it to encourage his readers to witness the performance of America's leading phrenologist.[37]

There were several other articles covering phrenology in his paper, but Douglass himself offered his most extensive statement about the science in an 1854 speech at Case Western Reserve. "The Claims of the Negro, Ethnologically Considered" was the first commencement address Douglass ever gave. And instead of hitting the graduates with platitudes about their futures, he decided to speak on the "natural history of man" from the perspective of various new sciences, focusing specifically on what they meant for the lives of Black Americans. A big part of the speech involved a direct confrontation with a book published in 1839 that Douglass saw as troublesome, to say the least. The book was Samuel Morton's *Crania Americana*, which was at once a defense of polygenism—or the theory that human races have different origins and are not branches of the same lineage—and a catalogue of different kinds of heads from across the globe. Morton used the heads he assembled in his massive collection—purportedly the biggest in the world—to create a racial ranking system. Unsurprisingly, he put Caucasian heads at the top of the hierarchy and Black heads at the bottom. And because the races were actually different species, Black people could never improve to the level of white people, no matter what they did.

Douglass spent much of his Case Western address tearing apart Morton's book. Denouncing the author for "isolating the negro race" from other races of the world, he underscored Morton's "contempt for negroes," particularly in the way he represented them on the page. In sacrificing "what is true to what is popular," Douglass explained, Morton presented an image of Africans that reinforced what white people already thought of them. And he presented images of the other races in the same stereotypical way. This led to huge problems in what was considered "representative." As Douglass told the audience, if "a phrenologist, or naturalist undertakes to represent in portraits the differences between the two races—the negro and the European—he will invariably present the *highest* type of the European, and the *lowest* type of the negro."[38]

Douglass was keenly aware of the problems and possibilities of representation in nineteenth-century America. The art and science of photography had recently emerged, forever changing the way people were represented in images that could be reproduced and shared broadly. Douglass knew, perhaps better than anyone else at the time, the power of photographs to capture likeness and to allow people to see what they wanted to see. That was one reason he sat time and again to have his portrait taken, making him the most photographed American in the nineteenth century. How he appeared in photographs, he knew, would be a big part of his influence and legacy.[39]

That's why he was so concerned about the way Morton and other scientists represented different races in their books. Given the power of imagery to shape people's understanding of the world, Morton's choice to pit the "lowest" example of a Black head against the "highest" example of a white head could have disastrous consequences for entire races. Phrenology, craniology, and other new sciences too often reinforced problematic representations with little concern for the harm they could do.

In making this astute argument, Douglass didn't say anything about phrenology being right or wrong. It could be used in problematic ways, to be sure, but that didn't mean the science itself was false. In the hands of people like Morton, who viewed races as fundamentally different, phrenology pointed to human divisions that could never be surmounted. But that's not where the science went in the hands of the Fowlers, who spoke of the unity of the human family and the ceaseless promise of improvement. And Douglass seemed to suggest as much in his Case Western address when he pointed to the work of George Combe as a model of how to properly represent Black people in science. Combe's *The Constitution of Man*, published in 1828, was a "great work" that held promise for science and the human race writ large, Douglass announced to his audience. In Combe's hands, much as in the Fowlers' hands, phrenology did right by the African "race."

No doubt when Douglass praised Combe in his Case Western speech, he was thinking back to 1846 when he had breakfast with the world-famous phrenologist. Douglass and Garrison were traveling together in the United Kingdom that year and stopped in Edinburgh. Combe invited the two men to dine with him in the morning and ended up talking their ears off. Despite his difficulty getting a word in edgewise, Douglass came away impressed with Combe and his knowledge of humanity. Praising his breakfast companion as an "eminent mental philosopher," Douglass noted how "Phrenology explained everything to him, from the finite to the infinite. I look back with much satisfaction to the morning spent with this singularly clear-headed man."[40]

Of course, even here Douglass never came out and praised phrenology directly. He praised Combe's love of phrenology, but he said nothing of his own feelings. In that way, Douglass shrewdly withheld final judgment on the science, publicizing it when he saw fit, rebuking it when it improperly portrayed his race, and always thinking of its long-term impact. This turned out to be a prescient decision. Douglass didn't know it at the time, but significant cracks were forming in the Fowler empire. And when the war came, everything changed.

CHAPTER TEN

DECLINE OF AN EMPIRE

THE CHANGE WAS SMALL AND SUBTLE, BUT IT REPRESENTED a monumental shift in the direction of practical phrenology: In 1855, "Fowlers and Wells" became "Fowler and Wells." The reason? Orson was going his own way, leaving Lorenzo the only Fowler at the helm.

In truth, Orson had had one foot out of the empire for several years. His entrepreneurial energy had shifted to architecture, and his octagon revolution seemed to be marching confidently forward (at least at that time). He was still, to be sure, fully invested in spreading the gospel of phrenology, but there was increasing distance from the others. Lorenzo, Charlotte, and their spouses lived in the same mansion and worked closely together every day. Orson lived upstate and regularly traveled the country to deliver lectures and to feel heads. Becoming independent would mean a better, more consistent schedule for himself; he planned to spend ten months out of the year lecturing and writing about phrenology, and the other two months—summer months—working on his property in Fishkill.

There was also real tension behind the scenes. For over a decade Orson had been addressing an issue that he deemed central to human life and improvement—sex—and courting controversy. In 1841, for instance, he rented the lecture room at Rutgers Seminary to speak on the topic of marriage. Before he even went on stage, local authorities warned that if his lecture addressed "the production of fine children," they would shut off the gaslights and throw Orson in prison for violating obscenity statutes. Not one to back down when humanity was at stake, he "dared them" to take action—which they didn't—and "continued to dare their kindred" over the next several years.[1]

Ever the businessman, Samuel Wells was increasingly nervous about Orson's sex talk. Concerned about the firm's bottom line, he worried that Orson's advocacy would not only hurt sales but put a legal target on their back. Orson thought Wells was acting cowardly and failing to support the firm's goal of spreading nature's truths. Then everything came to a head in the early 1850s when Orson and Lorenzo asked Horace Mann, the great education reformer, to write a treatise on masturbation. Mann wanted no part of the project, given that the subject "had ruined the reputations of all who had ever broached it." Nonetheless, in conversations about the treatise, it became clear to Orson that Wells "stoutly opposed" the idea and likely would have blocked the treatise from going forward if Mann had agreed to write it. Wells's opposition seemed to push Orson over the edge: "We *dissolved*" Fowlers and Wells, he later recalled, because Wells wanted to put business over truth.[2]

As a newly independent agent of truth, Orson was free to head in whatever direction the spirit of nature moved him. Accompanied by a chest of skulls, busts, casts, and other artifacts, he embarked on a lecture tour of Michigan, Ohio, Indiana, Illinois, Wisconsin, Iowa, and Minnesota. In each city he visited he stayed for a week or two in order to lecture and examine people's heads, which was the real profit center of his operation. Then, after spending two months back at home over the summer, he hit the road again, this time starting his lecture tour in Canada before dropping down to many of the same U.S. states he had visited the previous year. He also made appearances in Missouri, Kentucky, and Tennessee.

As a lecturer and practical phrenologist, wrote one paper in 1857, Orson remained a "master of his profession, and impresses most favorably all who have seen or heard him." Yet the very issue that Wells had worried about—controversy over the topic of sex—swirled around Orson's performances. In addition to lecturing publicly to broad, mixed audiences of men and women, he delivered private lectures to single-sex audiences—men only at one time, women only at another. The lectures covered "male and female perfection, illustrated by models," which involved some graphic details. But that was the point—to address delicate subjects that other people were too nervous or squeamish to tackle.[3]

Orson referred to this part of his work as "Sexual Science," as it covered "the Female Functions" and contained "truths of greatest practical importance to men, married and unmarried." In advertising his private lectures "exclusively to ladies," he pledged that "the knowledge imparted" would be "of inestimable value." When Orson stopped in Indiana in the spring of 1857, *The Evansville Daily Journal* felt the need to insist that lecture-goers attend the right performance for their sensibilities. While Orson's public lectures were "strictly chaste in every respect," containing nothing "in the least degree offensive to the most fastidious," his private lectures "may or may not be suitable, as they are not designed for promiscuous audiences."[4]

This was exactly the kind of response Orson hoped to cultivate as an independent phrenologist. He was a proper scientific gentleman in mixed company, but he was blunt, direct, and detail-oriented in his private lectures. In that way he could share truth with people who desperately needed it. "Let me help the 'bulls' of progress lift my race out of that 'old-fogy' slough in which they have been mired for ages," he wrote in his 1859 book *The Family*, which he self-published. "Allowed to choose my own name, and have it *true*, it would be, *Nature's Apostle*."[5]

It was a fitting title—and not only because he hoped to speak for nature. It was fitting because he would soon discover, as Jesus's apostles had discovered centuries before, that serving a higher truth often means dire consequences on this plane of existence.

ORSON'S RESTLESSNESS ABOUT business as usual at the Fowler empire must have been contagious, because a couple years after he left the firm and embarked on a new phase of his career, Lorenzo did the same thing. His exit started with a routine lecture tour. In January 1858, both he and Samuel Wells boarded a steamer bound for Charleston, South Carolina. Leaving the publishing department under Charlotte's watchful eye and the Phrenological Cabinet in the dexterous hands of Nelson Sizer, whom they had called back from the Philadelphia branch to manage affairs in New York, Lorenzo and Samuel traversed the nation to lecture and examine heads.

They began their venture in the Carolinas, where they stayed for a couple months, before heading farther down the Eastern Seaboard. Samuel worked primarily as Lorenzo's partner and agent on the road, overseeing publicity and securing lecture venues. That year, despite the rising sectional tensions between the North and the South, the two Yankees spent lots of time below the Mason–Dixon line, speaking in Montgomery, Alabama; New Orleans, Louisiana; Jackson, Mississippi; Memphis, Tennessee; and more.[6]

After months in the South they caught a steamer up to Nova Scotia, Canada, to deliver several lectures in a more temperate climate. Then in late 1858 they dropped back into the United States to make appearances in Pennsylvania and into the Midwest. In the summer of 1859, they went back to the Northeast and into Canada for several lectures, before stopping back home in New York for the holidays. For the first half of 1860, they continued to travel around the United States, staying mostly on the East Coast to speak, decipher characters, and unlock the secrets of the mind.[7]

In the summer of 1860, both Lorenzo and Samuel were back in New York (just in time for Lorenzo to examine the head of the infamous pirate Albert Hicks), and they devised a new plan. Looking across the Atlantic, they saw a massive opportunity. Even though the United Kingdom had been central to the early years of phrenology, it had not experienced the same kind of popular, practical surge the Fowlers had led in the United States. Phrenology overseas was more scientific and experimental than grounded in hope, progress, and improvement. What's more, George Combe, Europe's leading phrenologist, had passed away in 1858, leaving a bit of a vacuum in his wake. Lorenzo and Samuel were ready to step into that vacuum and foment the kind of practical-phrenological revolution in the United Kingdom that had spread across America decades before.

Lydia Fowler heard what they had planned and wanted in. For the past several years she had been practicing medicine in New York City and working on various reform efforts. Now, as civil war loomed on the horizon, she felt the same need for something new that had pushed her husband and Orson into new phases of their careers. So in the summer of 1860, Lydia, Lorenzo, and Samuel boarded a ship to Liverpool.

A month after their arrival, Lorenzo was already delivering lectures and spreading the good news of practical phrenology. He toured across England, Ireland, and Scotland, creating something of a sensation in the British papers. *The Manchester Courier* noted with pleasure Lorenzo's ability to accurately describe the character of people selected by the audience and encouraged "young and old to attend these lectures, for the young will learn to know themselves, while the aged will receive valuable hints how best to train those committed to their care." In Liverpool, the local paper noted that Lorenzo spoke "in a popular style" vivified with "considerable humor." The result was a hall "crowded with a most respectable audience" who "heartily applauded" the speaker throughout the lecture. In Sheffield, the local paper stressed the utility of practical phrenology: "To merchants in pursuit of business, to parents in rearing a family, and to man's thorough understanding of man, phrenology was declared to be invaluable, and its study earnestly enjoined." These were much the same responses they had received in the United States decades before.[8]

In the summer of 1862, Lorenzo, Lydia, and Samuel returned to the United States so that Samuel could take his place at the head of the publishing firm alongside Charlotte while Lorenzo and Lydia could pack up their household and move permanently to London. Hoping to replicate what they had done in the United States, they created a new phrenological cabinet, this time located on Fleet Street near Ludgate Circus. Lorenzo also established a U.K. publishing company in conjunction with William Tweedie, an English devotee of phrenology who helped arrange many of Lorenzo's lectures. The tried-and-true formula for promoting phrenology—lectures, demonstrations, a hands-on cabinet, and private examinations—worked well, as a "chord of sympathy" quickly developed between the Fowler family and the people of England, according to one paper. "Good feeling, confidence, and mutual respect" convinced Lorenzo and Lydia to stay in London for good.[9]

FROM ONE PERSPECTIVE, the 1850s were a boon for the Fowlers. They opened new branches of their cabinet, demand for their lectures and services reached new heights, and the public appetite for their printed material spread further and farther than ever. But from another perspective, the decade shook the foundations of everything they built. The two leaders of the movement, the public faces of practical phrenology, stepped away from the empire to pursue new paths in life. From either perspective, the 1850s were years of incredible change, even transformation—and not just for the Fowlers. Perhaps the most tumultuous in the nation's history, the decade fundamentally altered both the Empire of Skulls and America itself. Neither would (or could) be the same thereafter.[10]

The era of reform, which began around 1830, promised a new future for the adolescent nation—a stronger, brighter, more just, equitable, and prosperous future. But after two decades of heroic efforts to make America live up to its founding ideals, the nation seemed to be getting worse, not better. For one, support for slavery only increased—despite (and also because of) the work of abolitionists. The "great slave power," as it was known among many Northerners, was growing stronger with each passing year, as slaveholders organized themselves, partnered with politicians, and secured legal victories to hang on to their so-called property. The Compromise of 1850, which enhanced the Fugitive Slave Law, was horrifying proof of slavery's expanding influence, as it demanded that officials across the country, not just in the South, assist in the capture of runaway slaves.[11]

Then, in 1854, Congress passed the Kansas–Nebraska Act, which opened the Western territories to "popular sovereignty." This meant that settlers in the territories could decide by popular vote whether to allow or prohibit slavery. Recognizing that numbers would determine slavery's future and perhaps shift the balance of power between slave and free states, pro- and antislavery settlers flooded into the territories and began fighting for their cause. Bloody battles between free-staters and so-called border ruffians began popping up across the region, culminating in 1856 in the Sacking of Lawrence, which saw a proslavery mob pillage the Kansas town, loot businesses, and torch buildings. In response, friend of the Fowlers John Brown Sr. led a band of abolitionists on a revenge mission. In the dead of night they dragged five men from their homes along Pottawatomie Creek and hacked them to death with swords, severing limbs, slicing through torsos, and staining the dirt red.[12]

Amid the blood of Kansas, violence came to the floor of the greatest deliberative body in the world. Believing his family and region had been maligned, Congressman Preston Brooks of South Carolina marched into the Senate chamber on May 22, 1856, and strode confidently up to Massachusetts Senator Charles Sumner, who was seated at his desk. Before Sumner could stand, Brooks swung his thick wooden cane and connected with Sumner's skull, sending the antislavery statesman to the ground, covered in blood. After several more blows, Brooks simply sauntered out of the Senate chamber. In the weeks and months that followed, both he and Sumner were celebrated as heroes in their respective regions—Brooks for the caning, Sumner for dramatizing the destructiveness of the slave power.[13]

Then things went from bad to worse. In 1857, the U.S. Supreme Court handed down its decision in the Dred Scott case, ruling that people of African ancestry could never be U.S. citizens and that the government could not prohibit slavery in the territories. Convinced that slavery would never be destroyed but through bloodshed, John Brown Sr. penned a manifesto in the summer of 1859 that concluded with the ominous words, "Hung be the heavens in scarlet."

Now on a mission from God, Brown led a band of about twenty men on a raid of the federal arsenal at Harpers Ferry, Virginia. The plan was to secure weapons and supplies from the arsenal and to foment a slave uprising that would engulf the nation, finally bringing down the evil system. It failed miserably. Sixteen people were killed during the raid, including ten in Brown's party—two of them his sons. Six weeks later, John Brown was tried, convicted, and hanged for treason.[14]

After Harpers Ferry, civil war seemed not only likely but inevitable. Then the election of 1860 sealed the deal. When the supposedly radical Republican Abraham Lincoln was elevated to the highest office in the land, Southern states saw only one real option—create their own nation free from the tyranny of the North and built on the divinely sanctioned institution of slavery. A month after the election, South Carolina seceded from the Union, promptly followed by six other Southern states. Together they formed the Confederate States of America and elected Senator Jefferson Davis of Mississippi as their president. On April 12, 1861, Confederate troops opened fire on Fort Sumter in South Carolina and forced Union troops to surrender, claiming their first prize of the conflict. The Civil War had begun.

Many onlookers, especially in the North, figured the war would be over soon. Union forces were better trained and better equipped, and they could rely on the industrial machinery of their cities to crank out new supplies and weaponry. Predictions of a quick conflict almost came true in the spring of 1862 when Union General George McClellan surrounded the capital of the Confederacy, Richmond, Virginia, with more than 100,000 troops, far outnumbering Confederate forces. But instead of taking Richmond and likely ending the war, McClellan hesitated, believing, despite scouting reports, that Confederate forces in the city were stronger than they actually were. In the midst of McClellan's hesitation, Confederate General Joseph Johnston was able to escape and attack Union forces from the south, after which point Robert E. Lee assumed command of the Army of Northern Virginia. It was not going to be a short war.[15]

Casualties began to grow to unimaginable proportions. At the First Battle of Bull Run in July 1861, Americans were horrified by the almost 5,000 people killed, injured, or missing; but a year later, at the Second Battle of Bull Run, the number of casualties reached over 22,000. The deadliest single day in American history came on September 17, 1862, at the Battle of Antietam, when the number soared to over 22,700, showing just how brutal the conflict could be. Then in 1863, the bloodshed reached a horrifying pinnacle. Chancellorsville saw over 24,000 casualties, Chickamauga over 34,000, and Gettysburg over 51,000. By the end of the war, over 620,000 Americans were dead, more than the total of all previous American wars combined.[16]

This was the culmination of the era of reform. This was the result of the many efforts to make society stronger, healthier, happier, more just, and more equitable. Abolitionism, women's suffrage, education reform, prison reform, religious awakenings, communal living, and of course practical phrenology—they all ran headfirst into more death than the nation had ever known. The torrents of life that stained America from 1861 to 1864 were, in many ways, a repudiation of everything the Fowlers and other progressive advocates had worked to achieve. The hopeful adolescent nation grew up quickly into a jaded adult.

But the horrifying number of casualties was only one factor in the ongoing upheaval. The Civil War followed closely on the heels of the advent of photography, making it the first major conflict in American history to be documented visually. The famous Civil War photographer Mathew Brady and his associates brilliantly captured the death and destruction through their lenses. Because of the state of technology at the time, they were unable to capture combat itself, but they painstakingly documented the aftermath of several battles. Bodies strewn across fields; friends and neighbors lying cold along fence lines; soldiers decaying with muskets propped on top of them; nameless countrymen prostrate over rocks; mass graves packed with rotting corpses and festering maggots—these were the images that brought the war into vivid relief for noncombatants.[17]

African American soldiers cleaning up the dead in Cold Harbor, Virginia, a year after the battle

Brady's photography gallery in New York displayed many of the images he captured, including from the bloody Battle of Antietam in 1862. One of Brady's associates even turned battlefield images into collectible cards that people could view, often in stereoscopic 3D, at home. While newspapers and periodicals were not able to print the photographs directly, given the state of printing technology, they used the images to make engravings that were then reproduced across the country. Unlike ever before, Americans were face-to-face with the horrors of war. Oliver Wendell Holmes Sr., the famous physician and essayist who visited a battlefield to search for his injured son—the future Supreme Court Justice Oliver Wendell Holmes Jr.—wrote for *The Atlantic Monthly* about how war photographs brought him back to what he had seen in person: Viewing the photos "was so nearly like visiting the battlefield to look over these views, that all the emotions excited by the actual sight of the stained and sordid scene, strewed with rags and wrecks, came back to us, and we buried them in the recesses of our cabinet as we would have buried the mutilated remains of the dead they too vividly represented."[18]

Sometimes photographers arrived on battlefields long after the carnage, and their cameras captured not dead bodies but piles of bones. In one particularly arresting image, African American soldiers cleaned up from the Battle of Cold Harbor a full year after it took place. One of the soldiers posed for the camera by crouching alongside a stretcher full of sun-bleached bones, tattered clothes, and knowing skulls. Another image showed the aftermath of the Battle of the Wilderness in Virginia, featuring a half dozen human skulls littered in a field like unremarkable animal bones. One skull rested atop a small tree trunk—a warning to future generations about the depravity of man.

These were the skulls that gained prominence in the minds of many Americans, and they were anything but scientific. The Fowlers had spent the previous twenty-five years proclaiming that practical phrenology meant improvement for individuals, communities, and nations. It provided a natural foundation to the various reform efforts then underway, all designed to put America on the right track. Yet the unthinkable violence of the war and the associated photographic evidence told a story about how futile those reform efforts and theories of improvement had been. Whereas the skulls of the Fowler empire had promised progress, the skulls of the 1860s brought people back to reality, reminding them of the evil of humanity.

WITH LORENZO AND Lydia now living in London, Charlotte and Samuel were left trying to keep the New York operation afloat during the tumult of the war years. And they did an admirable job. In response to President Lincoln's 1862 call for 300,000 volunteers to fight, they gave their Cabinet over to the firm's engraver, William Howland, to muster the 127th regiment, which marched to Virginia in October of that year. Many longtime staff members also enlisted in the fight.[19]

During the war Fowler and Wells publications regularly featured news from the front lines and various war-related topics. Wells believed it his duty as editor of *The Phrenological Journal*, and the duty of all periodical editors, to focus attention on the Union cause and to save other matters, when possible, for the time when "peace and business activity shall have returned." He and Charlotte published articles on military terminology, the various ranks of enlisted men, and soldier rations. They covered the weaponry being used on the battlefield and the numerous patents and inventions emerging from the conflict.[20]

They also printed phrenological descriptions of pivotal figures in the war. According to the journal, General George McClellan, head of the Union army, "seldom feels the necessity of asking advice, because he perceives instantly the circumstances by which he is surrounded, and his first judgment is his best." General Ormsby Mitchel had a prominent middle forehead, accounting for "a great command of facts, geographical talent, excellent memory of time, and a tendency to punctuality." They prominently featured an article about William Tillman, a Black man who recaptured the schooner *S. J. Waring* from pirates during the war. His head, which they examined in person, featured "uncommonly large Perceptive organs, which give practical talent and good common sense." Brigadier General Ambrose Burnside had a brain "large in the thinking department" and "an excellent power to plan," which made him a great military leader. Of course, the journal proudly featured a phrenological description of Abraham Lincoln, whom Wells described as "true to his higher nature" and "governed by moral principle rather than by policy." "If not the best man for the situation," Wells questioned, "where can you find a better?"[21]

Unsurprisingly, sales at Fowler and Wells took a hit during the war. Realizing that he lacked the number of agents needed to sell subscriptions and to hawk publications, Wells called on women, the ones left to manage the home front, to sign up as agents for the firm, given that their "increased responsibilities" necessitate, more than ever before, "the aid which the *Phrenological Journal* would give her." He also pleaded for faithful readers to make "a special effort to send us new subscribers," given that "nearly every kind of publication in wartime languishes." Nevertheless, sales continued to decline, and

A Repository of Science, Literature, General Intelligence.

VOL. XXXIV. NO. 3.] NEW YORK, SEPTEMBER, 1861. [WHOLE NUMBER, 273.

Published by
FOWLER AND WELLS,
No. 308 Broadway, New York.
AT ONE DOLLAR A YEAR, IN ADVANCE.

Contents.

MAJ.-GEN. GEORGE B. McCLELLAN.

PHRENOLOGICAL CHARACTER AND BIOGRAPHY

PHRENOLOGICAL CHARACTER.

THIS portrait, presented for our examination by a friend, without any indication of the name of the original (nor had the examiner ever before seen a likeness), indicates the following qualities: In the first place he has a most marked and positive Temperament, evincing activity and endurance in a high degree. His phrenological developments are also marked. His Perceptives are large; hence his mind is quick, clear, and practical. He grasps the facts and conditions of things almost instantaneously, and forms a judgment respecting them with uncommon rapidity, clearness, and accuracy. He seldom feels the necessity of asking advice, because he perceives instantly the circumstances by which he is surrounded, and his first judgment is his best. He is remarkable for order, for precision, and for mathematical accuracy in all he does. His head is broad, evincing uncommon force, courage, fortitude, and self-reliance; he never felt the necessity of being helped, protected, or sustained. He has Cautiousness and Secretiveness large enough to give him policy and prudence; and his Destructiveness and Combativeness, joined with very large Firmness, give him that self-dependence and consciousness of power which gives promptness to his decisions and earnest execution of his plans.

MAJ.-GENERAL GEORGE B. McCLELLAN.

His large Form, Size, Locality, and Individuality give him great talent to observe, sketch, to carry a picture or outline of things in his mind, and to remember geography, local position, and adjustment. These are very important qualities in an engineer or military leader. His Constructiveness, Form, Size, Order, and Calculation being large, qualify him for engineering, mechanism, and for forming combinations and inventions.

His Causality is large; hence he has an inquiring mind, is fond of investigating, and learning the philosophy of everything. He is capable of looking ahead, and seeing the end from the beginning; and his very large Perceptive qualities tend to open his pathway, and to throw light on everything in his immediate vicinity. He will perceive the best way and most ready access to results; and if he were thrown into straits of difficulty, he would form new plans and combinations almost instantaneously. Self-Esteem, Firmness, and the executive faculties lay the foundation for uncommon independence of spirit, and a desire to pursue

The September 1861 issue of The American Phrenological Journal *detailing the phrenological character of Union General George McClellan*

the firm, like so much else in the nation, would never be the same once the conflict was finally over.[22]

To mitigate the decline once the war was over, Charlotte and Samuel focused increasingly on the bottom line and establishing a new brand identity. They changed the name of the firm to S. R. Wells, pushing the Fowler name officially out of the operation. Improvement remained at the core of what they did, but Wells himself dreamed of a more generalist publishing operation that could adapt to demographic shifts, political upheavals, and the wonders of the age.

Hoping to expand their market, Charlotte and Samuel launched a number of new titles, including *Hand-Book for Home Improvement, How to Write, How to Talk, How to Behave, and How to Do Business* and *The Farm: A Pocket Manual.* There were also pamphlets on various social issues, including marriage, family life, and women's place in the modern world. They even created a new periodical called *The Science of Health*, which, as the cover page explained, focused on "light, air, temperature, electricity, diet, bathing, sleep, exercise, and rest" and was "amply illustrated" for a reading public who loved visuals. In time, *The Science of Health* would merge with *The American Phrenological Journal* in order to cut costs, resulting in *The Phrenological Journal and Science of Health.*

Then came 1873 and a year of disasters. It started out with a "terrible fire" that engulfed the firm's book bindery and led to delays for several publications, including *The Phrenological Journal.* This was not the first fire the firm had endured, nor the first delay in shipping, but it was a harbinger of what was to come. Later in the year the nation was rocked by a financial panic that destroyed countless businesses, battered family finances, and upended American society. At least one hundred banks across the country failed, dozens of railroad companies had to shutter, and numerous construction projects came to a halt. Many Americans were out of work, had their wages cut, and lost their homes.[23]

When the panic hit, Charlotte and Samuel had to make a number of adjustments to keep the business going, and they succeeded, albeit with some big changes. In 1875, real estate speculators started

Exterior of yet another phrenological cabinet in the declining years of the Fowler empire

buying up buildings in the neighborhood of the Phrenological Cabinet and S. R. Wells publishing. Eventually "the money changers secured the property," Wells told readers of the journal, and they had to vacate their office. But they found a new space at 737 Broadway, just a mile up the road, surrounded by all the big publishers in the city, and next to the Astor Library. Wells was confident, or at least projected confidence, that getting kicked out of their old building was a blessing in disguise, as it gave them the opportunity to operate alongside the major players in the nation's literary world.[24]

Charlotte and Samuel together oversaw the move to the new location, which included elaborate plans to transport thousands of delicate artifacts and stacks upon stacks of publications and supplies.

It was a bigger job than either of them had anticipated, and it took a toll. In early April 1875, Wells was working with painters, carpenters, and designers to get the new office ready when he caught a cold, lost his energy, and had to go home for rest. Everyone on the job site figured he had simply overworked himself and needed a few days to recuperate. But Wells had contracted pneumonia, which he fought off for ten days, only to lose the battle on April 13. "The pale messenger came for the spirit of him," Charlotte wrote in announcing his death at the young age of fifty-five.[25]

Condolences poured into the office, many of them published so the world could see the influence Wells had had on humanity. Charlotte herself received countless messages, including from Susan B. Anthony, who noted how despondent she became at knowing "that there is no Mr. Wells" by Charlotte's side.[26]

With Wells's passing, control of the firm passed to Charlotte, while Nelson Sizer continued running the Cabinet and leading the head examinations. To recognize this new phase of the business, Charlotte changed the name of the firm from "S. R. Wells" to "S. R. Wells & Co." Adept as a business leader, she kept the firm moving forward and attracted considerable attention for her efforts. In 1878, several newspapers ran a story on Charlotte, her tireless work ethic and brilliance leading an established operation. Calling her "sole proprietor and business manager," *The Milwaukee Daily News* noted that under her direction *The Phrenological Journal* had "come out victorious through storms which have sunk so many more formidable enterprises." In the trying times of the 1870s, "her clear brain and indomitable perseverance were the central power which from the very first ensured success."[27]

EVEN THOUGH HE had moved on from the firm he helped to create, Orson still saw tremendous promise for practical phrenology. In fact, he was more bullish than ever about its transformative power. In 1863, after bouncing around the country as a lecturer, and after the unfortunate demise of his octagon revolution, he settled in

Manchester, Massachusetts, and opened an independent phrenological office in Boston. Because Fowler and Wells still had a branch of the Phrenological Cabinet in that city, he was now a direct competitor with his sister and Samuel Wells. Tragedy struck the following year when Orson's wife, Eliza, passed away at the age of fifty-eight. One year later, in 1865, he remarried, joining hands with Mary Poole. Like many men at the time, Orson did not want to stay single for long.[28]

The newlywed couple settled into their new life in Massachusetts, and Orson settled into a new phase of work, dedicating himself to the role of Nature's Apostle more than ever before. His daily schedule was to rise before the sun, write until midmorning , have breakfast, examine heads, read proofs of his various publications in process, and answer correspondence. Then he'd rest until seven, at which point he would write again until eleven at night. He'd do it all over again the next day.[29]

All this work was part of his most ambitious book project to date, a scientific treatise that he called a "complete work on *all the departments* of both man's mind and body," uniting "*all* branches of Anthropology into one collective *whole*." After years of toil, his gift to humanity appeared in print in 1870, and it consisted of two volumes of over one thousand pages each. A third one-thousand-page volume would follow shortly thereafter.[30]

The first book was called *Human Science*, but its full title signaled the ambition of the project: *Human Science, or Phrenology: Its Principles, Proofs, Faculties, Organs, Temperaments, Combinations, Conditions, Teachings, Philosophies, Etc., Etc.; as Applied to Health, Its Value, Laws, Functions, Organs, Means, Preservation, Restoration, Etc.; Mental Philosophy, Human and Self Improvement, Civilization, Home, Country, Commerce Rights, Duties, Ethics, Etc.; God, His Existence, Attributes, Laws, Worship, Natural Theology, Etc.; Immortality, Its Evidences, Conditions, Relations to Time, Rewards, Punishments, Sin, Faith, Prayer, Etc.; Intellect, Memory, Juvenile and Self Education, Literature, Mental Discipline, the Senses, Sciences, Arts, Avocations, a Perfect Life, Etc., Etc.*, by Prof. O. S. Fowler. Effectively, the book contained everything about the world that Orson had learned over his long career. Some sections were direct cut-and-paste jobs from previous publications. Others were rearrangements, reworkings, and restatements, all in the service of ultimate, complete truth.

The second one-thousand-page volume published in 1870 was titled *Sexual Science; including Manhood, Womanhood, and Their Mutual Interrelations*. This book contained Orson's lectures on sex education combined with sections of previous books, including *Matrimony, Love and Parentage*, *Hereditary Descent*, and *Offspring*. *Sexual Science* tackled reproduction proudly and boldly, as Orson aimed to remove all prudery from public discussions of sex. He hoped to make his readers "literally tremble in view of their past sexual errors and present dilapidations, till they imploringly inquire, 'How can I be saved' therefrom? And to teach all how to carry their sexual perfection and enjoyments up to the highest attainable point." The book, he boasted, "makes no apologies. It asks no favors. It submits its claims to public attention on its naked merits, and appeals to the good sense, good taste, and self-interest of mankind."[31]

Shortly after publication, Orson reported that *Sexual Science* was "a veritable God-send" to "all wives, all maidens," and that "one hundred thousand" of them rate the book "next to their Bibles" in importance for family development. Setting aside the fact that Orson no doubt concocted that number, the book made a splash in the marketplace. "Every single reader" of it, he reflected, "who has attempted to put its teachings into practice, has become perfectly enthusiastic over their *results* as experienced in each of their cases." Perhaps the sex-education revolution would happen just as Orson wanted.[32]

Or perhaps not. For some reason—and the historical record is unclear exactly as to why—he quickly pivoted from *Sexual Science* to another one-thousand-page book on the same subject. *Creative and Sexual Science* was his second attempt to summarize everything he knew about sex and human nature, and he started working on it as soon as *Sexual Science* appeared in print. *Creative and Sexual Science* actually contained two copyright dates—1870 and 1875—and was also brought to market by a different publisher. It's quite possible that the publisher of the first sex book got cold feet as soon as it was printed, leaving Orson to find a new publisher who would let him write even more boldly and comprehensively in the second book.[33]

The new book, Orson recounted, was developed with "double the mental labor" as *Sexual Science* and included "one-fourth more subjects," along with "three-fifths more illustrative engravings." More importantly, Orson was able to incorporate the delicate subjects he had omitted from the first book for fear of assaulting public sensibilities. This time around he would omit nothing "on the score of propriety."[34] Emboldened to make the title of Nature's Apostle include unafraid sex educator, he published a book that was truly out of the ordinary for its time, describing in graphic detail what was not even to be whispered about in public. He explored the natural laws of attraction, including how women love "sexual vigor and passion in men," and how the entire world exalts female beauty. Men, he related, are particularly attracted to women whose bodies highlight their essential female function—that is, bearing children. A large pelvis, he wrote, is the great beautifier of women, as men greatly admire a prominent "mount of love"—or "a large and well formed pubis." Men also cherish "large, luscious bosoms," which, aside from the pubis, are the most attractive physical feature of female bodies.[35]

Once again playing the part of Victorian moralist, Orson made proper courtship a central issue in the book. Men, he wrote, ultimately want to impregnate women and women want to be impregnated by men, but during courtship this innate desire creates huge sexual tension that can derail people's lives. Women who permit "sexual freedoms" will find themselves used, then cast aside by cocksure seduction artists, who deserve to be "horsewhipped" for sullying young women. "Yielding girls be forewarned," he declared, "that all courting liberties both gratify your fellow without marriage, and disgust him with you, and you of him."[36] To set courtship on the proper course, Orson dedicated hundreds of pages of the book to the lost art of lovemaking. As he used the term, lovemaking was not equivalent to sexual intercourse; it referred rather to all the interactions, evaluations, assessments, emotions, habits, and feelings that go into two people joining together inside and outside the bedroom. Of course, sex was part of lovemaking, and proper lovemaking required good sex. "Unsatisfactory intercourse kills the Love of both," he explained, and women "must feel at perfect ease and participate fully" in every sexual encounter. To ensure as much, he insisted, couples must regularly rekindle the fires of love and not let "amatory excitement" derail their union. This meant nurturing Love "at a stated *hour* each day" in order to create "conjugal excellence."[37]

Then Orson got into intercourse itself, which he described as "absolutely magical": "Every wife who fulfills this function right with her husband can lead him where and do with him just what she pleases; for his complete satisfaction here is precisely what constitutes her magic wand over him." In graphic detail he explained how penal erections are a perfect fit for the female anatomy: "The male penis is Nature's depositing instrument. Its rigidity becomes indispensable in order to make its way to the uteri, so as to deposit its seminal messenger of life there." Because of this perfect fit, intercourse "summons all the organs and parts of the system to its love-feast, *compels* their attendance, and then lashes up their action to the very highest possible pitch." Orson even dedicated an entire section of the book to sexual positions. He favored having the woman on top because it allowed for the fullest penetration, getting the penis as close to the womb as possible. He also championed what he called "Indian" style, which meant both partners lying on their sides. "Thank phrenology," he wrote, that these sexual lessons "have at last been unfolded; and employ and enjoy them in treating yourself to the richest amatory feast, and the highest sexual luxury, of which your remaining sexuality renders you capable."[38]

Because Orson believed that reproduction was the true and proper end of sex, he spent significant time decrying nonreproductive sexual acts. This included the act of "withdrawals"—or the man pulling out right before ejaculation—which he called "inherently vulgar and vulgarizing, repugnant and debasing to all purity and refinement." He had similarly harsh words for condoms, the use of which was "appallingly prevalent . . . Sexual depravity, where is thy limit?" Of course, the elephant in the room around nonreproductive sexual acts was masturbation, which he called the "worst of sexual vices," insisting that it "outrages nature's sexual ordinances more than any or all the other forms of sexual sin." The evil was almost universal in men, and shockingly widespread in women, and its effect ruinous. The ravages of masturbation, he warned prophetically, "follow and prey on you forever! You can never fully rid yourselves of the terrible evils it inflicts. You may almost as well die outright as thus pollute yourselves."[39]

Orson also set his sights on a new affront to nature—at least new to him, given what he had written in his 1843 book *Hereditary Descent*. For some unknown reason, he now spoke out against race mixing, which he called "forbidden by Nature" and something not to be "perpetrated by man." It's unclear why Orson reversed course from his striking, radical position three decades before. Earlier he had made the point that race mixing was a boon for humanity, given that it facilitated the emergence of new, heretofore unknown traits. But toward the end of his life, his views on race and sex regressed to become in line with the prevailing position of the day.[40]

Nevertheless, Orson heralded this new book as a masterpiece, as a true gift to humanity. Calling it one of the top books "of this century, *or any other*," he beckoned readers to spread his gospel of sexual salvation far and wide. Then he offered a benediction akin to what churchgoers would hear in the pews on Sunday morning:

> May this book perfect the race by unfolding creative and sexual science; revealing the natural laws and facts of Love; guiding it upon right object; promoting marriage and conjugal felicity; showing how to create, carry, bring forth, and bring up many more and immeasurably better children, and telling all how to educe their own and children's manhood and womanhood into perfect husbands, wives, and descendants. God bless you all.

With those parting words, Orson believed he would usher in a new era of sexual, mental, and spiritual health. What actually happened would send Orson to an unmarked grave in the Bronx at the ripe old age of seventy-seven.[41]

IN 1880, CHARLOTTE made a bold move. Now without her soulmate, and at the helm of a once-successful publishing firm, she knew she needed to reinvigorate her business. The best way to do so, she concluded after thinking the matter over, was with a kind of homecoming. Through "a cordial harmony of interests," she announced to readers of *The Phrenological Journal*, Orson was rejoining the family and the firm. Deeming it "expedient to combine the strength and influence of the old workers with those who are younger," Charlotte explained that Orson, "whose well-known name has been so intimately connected with Phrenology in America," would "resume his old place" and lend "his prolific pen" to the publisher once again. The "phrenological family of co-workers" was back together, Charlotte wrote excitedly, and "mankind needs that which we teach." Fittingly enough, the firm once again changed its name—back to Fowler and Wells, reflecting the stewardship of both Orson and Charlotte.[42]

For a brief window of time, the future looked brighter for practical phrenology than it had since before the war. Then Orson's past caught up with him. In 1881, Orson's second wife filed for divorce, claiming rampant infidelity on his part. He countersued her, claiming that she was also unfaithful and that he could "substantiate it with half a dozen individuals" whom he knew by name. Some newspapers at the time reported that his soon-to-be ex-wife had been bilking him out of his earnings for several years and, now in the divorce suit, was seizing "his books, his plates, his busts, his copyrights and everything," leaving the old man "with not a dollar left." As one paper joked, the professor, a supposed expert on "connubial happiness," was "too busy looking after the welfare of his numerous clientage to attend properly to his own" marriage.[43]

But that was only the first problem in what became a horrible year for Orson. The previous summer, Orson had split from his partner and business manager, Max Bachert, who had assisted the famous phrenologist for years. For a while, their parting had seemed amicable. But unbeknownst to Orson, Bachert was in possession of documentary evidence of Orson's sex-education hypocrisy. The old man's "private lectures to ladies" were apparently much more than lectures, often taking on "an immoral character" and becoming "grossly obscene in action and speech."[44]

When word of Orson's divorce hit the newspapers, several journalists started asking questions about the allegations of infidelity, and they soon connected with Bachert, who was more than happy to share what he knew. One journalist, Nym Crinkle (real name Andrew Wheeler), worked with Bachert and learned "all sorts of hard things about old Prof. Fowler," leading one paper to joke, "Prof. Fowler is preparing to examine the head of Nym Crinkle by passing over it a sympathetic brick."[45]

Crinkle's reporting ended up in several issues of the *Chicago Tribune* and included accounts of Bachert's documentation. The journalist even showed the evidence of Orson's misdeeds to judges, politicians, and ministers, who promptly denounced what they saw as immoral and obscene. A New York minister, Howard Crosby, deemed Orson's public writings and private letters of "a nasty, filthy nature," and resolved to bring Orson to account for corrupting the morals of the youth. "The publication and circulation of books of this nature, and the writing of letters of the pernicious, disgusting, and immoral character of those shown to me," insisted the minister, "are calculated to do an injury to society, and especially the younger portion thereof, by sowing seeds of immorality and vice." It was a vile conspiracy, the *Tribune* wrote, in which Orson used the science of phrenology as a cover to debauch "the minds of young females" and to foment "a prostitution of the sexes under the guise of education."[46]

These revelations about Orson's lecherous activities ended up on the desk of Anthony Comstock, New York's infamous moral crusader, who had founded the New York Society for the Suppression of

Vice in 1873. Comstock supposedly viewed Bachert's evidence and concluded that Orson's writings could only "be classed among vicious and filthy literature." Several newspapers at the time reported that Comstock was preparing to shut down Orson's supposedly scientific sex cult.[47]

For whatever reason, that never happened, and Bachert's evidence never came to light. But Charlotte knew that the damage to practical phrenology and the Fowler name had already been done. In response to the allegations against her older brother, she published a circular denouncing him and denying knowledge of his sexual sins. She also lashed out at the "acrimonious and unwarranted attacks upon us, as prompted by unworthy and hostile motives."

Nym Crinkle wasn't buying it. Although Charlotte was "making some vigorous protests against the association of that old arch-lecher, O. S. Fowler," Crinkle wrote in the *Tribune*, the "simple truth" was that "for long years the rottenness of this phrenological syndicate has been so carefully guarded by the families (allied by marriage) having a monopoly of the so-called science that it has been kept from the public eye, but now that it is being exposed those who have much to fear are beginning to tremble, and are seeking to save themselves by denial of association with the one who has brought his disgrace upon them."[48]

In August, Bachert himself entered the fray by publishing his own circular, attacking Charlotte for turning a blind eye to what her brother was doing, and detailing everything he knew about Orson. After years of witnessing the old man's misdeeds, Bachert explained, he approached Orson and begged him to stop. "On every occasion Mr. O. S. Fowler promised to amend his conduct"; yet time and again he continued his "sexual intrigues" and refused to stop sinning. Bachert even discovered Orson "to be sustaining the most disreputable relationship with certain female quacks, and to be writing to them grossly immoral letters, which actually undertook to systematize sexual vice." Then Bachert made a gobsmacking claim: Most of the money that Orson earned over the years went to "hushing up the scandals" that would have otherwise come out. "At least ten thousand dollars of his money, most of which was made by my exertions, went to pay up old debts and wipe out old judgments created by the extravagance of his own relations."

Ultimately, Bachert insisted that he wasn't seeking money; he simply wanted the old man to stop his immoral pursuits and to retire from public life. This "professional scientific fraud and charlatan" preached "total abstinence all over the land" while "indulging daily in the finest wines," and his "claims of morality were supplemented with vices that would have shut him up in the penitentiary, if exposed and punished." Using phrenology as "an excuse" to get close to women and championing "morality only as a cloak," Orson spent his years "perverting the minds and attempting to debauch the bodies of his hundreds of dupes who happen to be of the opposite sex, and are not old enough to tell the difference between a savant and a sensualist."[49]

Orson never responded publicly to Bachert's accusations, the supposed evidence never materialized, and the authorities never intervened. So any firm conclusions about Orson's conduct would be hasty. But the damage to his reputation had been done, and the relationship with his siblings, especially Charlotte, was effectively destroyed.

The Fowler empire was crumbling, and not just because of Orson's sex scandal. Public perceptions about science, humanity, and the world were rapidly changing in post–Civil War America. The nation was growing up fast, right at the same time Orson, Lorenzo, and Charlotte were entering the twilight of their years.

EPILOGUE

THE HOPE OF IMPROVEMENT

IN 1873, A MAN OF MIDDLING HEIGHT AND VIVACIOUS HAIR passed by Lorenzo's phrenological depot on Fleet Street in London. Seeing the Fowler name over the door, he smiled as he recalled his upbringing in a small Missouri town and the electricity that surrounded a Fowler lecture. Because the townsfolk "read and studied" the publications of Fowlers and Wells with the excitement and dedication of truth-seekers, a visit by the brothers "gathered the people together" and revealed "the marvels of phrenology," after which Orson and Lorenzo "felt their bumps and made an estimate of the result, at twenty-five cents per head. I think the people were almost always satisfied with these translations of their characters."[1]

Recalling the Fowlers with nostalgia and amusement, the man decided, as many before him had, to make "a small test of phrenology." Hiding his identity with an assumed name, he entered the business and sat down as Lorenzo "fingered my head in an uninterested way," describing character traits "in a bored and monotonous voice." Apparently Lorenzo found conflicting aspects of the man's personality—large Combativeness but even larger Cautiousness—that seemed to cancel each other out. One thing he did find was a startling cavity across the man's Mirthfulness, which indicated "a total absence of the sense of humor." Lorenzo explained that "he often found bumps of humor which were so small that they were hardly noticeable, but that in his long experience this was the first time he had ever come across a *cavity* where that bump ought to be."

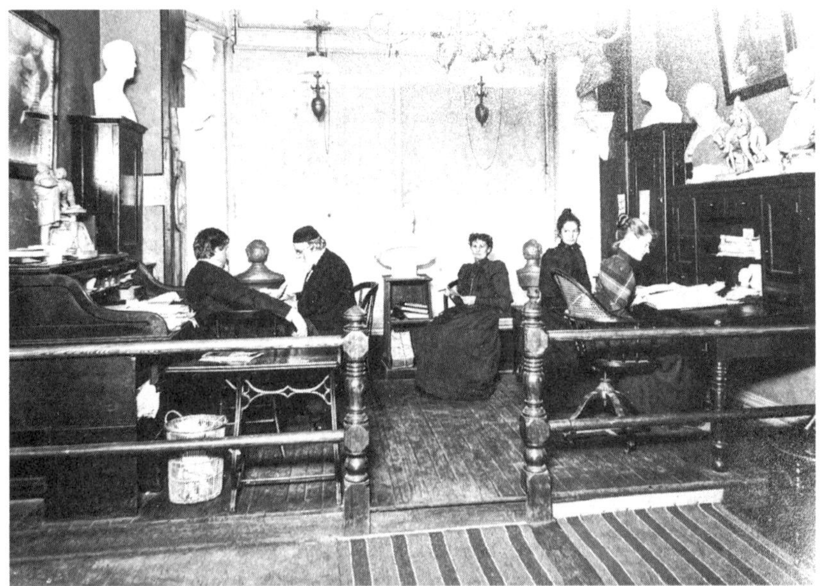

Interior view of the phrenological cabinet in London where Mark Twain received his examination from Lorenzo, who is seated wearing a skullcap

It was a strange sitting, but the man, smirking to himself, left with a chart and written description in hand.

After waiting three months so that Lorenzo would, hopefully, forget his face, the man returned to the depot—this time presenting his card upon arrival. Peering down at it, Lorenzo saw the man's name: Samuel Clemens, also known as Mark Twain. Thrilled with the celebrity in his office, he again examined Twain's cranium through his bushy hair, tracing the curvatures of the great American humorist. What he found now, at least according to Twain, was that the cavity over Mirthfulness "was gone, and in its place was a Mount Everest—figuratively speaking—31,000 feet high, the loftiest bump of humor he had ever encountered in his life-long experience!" With two contradictory charts, Twain left the office "prejudiced against phrenology," and particularly against Fowler for exploiting the science for personal gain.[2]

Because the original charts of Twain's test have not survived, his recollection is the only remaining account of Lorenzo's supposed failure. Nevertheless, Twain's experience was indicative of changing attitudes about the Fowlers and practical phrenology happening at the time. There had always been opponents trying to take them down, but in the 1870s the widespread belief in improvement, even perfectibility, that had captivated the American people for so long shifted toward the same skepticism that Twain felt, especially with the horrors of war so fresh in collective memory. It was time to face the world with a stern realism, not some swirling, airy faith in phrenological progress, or so the general consensus seemed to be.

What's more, science had come a long way since the Fowlers had begun their campaign to make practical phrenology the basis of everything. In the middle nineteenth century, U.S. higher education and scientific research underwent a transformation via the influence of German universities, which practiced narrow, highly specialized, discipline-specific scientific inquiry. Their understanding of the scientific method, which came to define science in the United States, had little room for the kind of experiments the Fowlers undertook. Science was exacting, rigorous, and replicable, while the Fowlers seemed to traffic only in personal charisma, flattery, and hope. In fact, the very idea of a single, universal, unifying science was laughable amid the increasing specialization and professionalization of the German model. Humans and the world were too complex for the kind of "science of everything" the Fowlers championed.[3]

Then there was the arrival of Darwinism. Darwin's research on evolution, published in 1859, changed people's thinking about human development. The Fowlers had long spoken of development and improvement, but they were focused on a small timescale—progress in just a few years or decades. Darwin's theory turned people's thinking toward development across history, where progress was measured in epochs, not in years, and where species were the controlling factor, not individual organs. The Fowlers had faith that individuals could check, counteract, and redirect what nature had given; under Darwinism, that idea was silly, which Darwin himself pointed out in one of his books.[4]

This steady turn away from practical phrenology caught up with the Fowlers. People continued to attend their lectures and buy their books, but not in the same numbers as before. There certainly wasn't the aura of world-changing science around their work that there had been in the past. After his sex scandal, Orson kept working as a practical phrenologist, lecturing, writing, and examining heads across the United States. He even found time for a third marriage. In 1882, a year after the scandal broke, he wed Abigail Ayres, who was thirty years his junior. She took on the role of Orson's publisher, business manager, and phrenologist in training.[5]

But Orson was getting old, and time was not on his side. In the first half of 1887, he went on a lecture tour as he had done for years, speaking to audiences in California, Arizona, and Texas. But at seventy-eight, he found the travel schedule more exhausting than ever. In August, after he had been home for a few weeks in Sharon, Connecticut, where he and Abbie lived, he caught a cold that debilitated him and caused severe "spinal trouble." After thirty hours of pain, Orson Fowler passed away on August 18, 1887.

Newspapers across the country took note of his passing, and their commentary was predictably uneven. One paper dismissed phrenology as a "shadowy science" with no real students anymore, though it lauded Orson's moxie for getting people excited about it. His "style of long hair and sleek ways," the paper maintained, was central to his success. Another paper was even more biting, noting that his passing "recalls the palmy days of the phrenological humbug" that was "so profitable to charlatans." "At one time Fowler & Wells were making more money than any other concern on Broadway," the paper noted, "but the rage for such things is now much abated, and phrenological lecturing does not pay."[6]

Other papers, however, remembered him as "a hard thinker, a most indefatigable worker," who "made his mark on his age and country." "Few public men of twenty-five years ago," wrote another paper, "but could relate some striking incident in connection with the examination of his head by one of the Fowler brothers, and it is safe to say that to-day there are few intelligent men who are not familiar with their names. O. S. Fowler was always foremost in all works of reform and progress." Still another paper recalled the "tidal wave of success" that followed him through life: In addition to "his fortune and his immense printing house, . . . his example remains the guidance of all. The secret of his success lay in his belief in himself and in his industry and tenacity of purpose."[7]

Yet even in the celebratory coverage of Orson's life, there was a sense that his passing marked the end of an era—the result of a slow, steady decline that had been happening for decades. One wire-service article, reprinted in numerous papers, remembered Orson as "The Prophet of Phrenology" and praised the tenacity and success of his "itinerant business." "The founder of practical phrenology in this country," he and his siblings made the science understandable to the masses and published "nothing not productive of good morals." Nevertheless, the article concluded, "The great interest which Mr. Fowler's enthusiasm created in phrenology has about died down, and many of the younger readers of these lines only know of it incidentally." For the amusement of those readers, the article printed an illustration of a phrenological head and its corresponding organs.[8]

Fortunately, coverage of Orson's passing did not mention the sex scandal of 1881. And even Orson's siblings, who had effectively disowned him, took the time to celebrate his life. Charlotte's periodical *The Phrenological Journal* praised his "rare independence" and "spirit of a ready, thorough-going worker." It was Orson who brought the lofty theories of European phrenologists "down to the comprehension of the ordinary, every-day intellect, and gave them point and application by well chosen examples from any company that was assembled to hear him." Lorenzo's London-based *Phrenological Magazine* wrote, "The science of phrenology has lost one of its most intelligent and enthusiastic advocates, and New York a useful citizen, in the death of Professor Orson S. Fowler." For decades, the periodical continued, Orson labored "earnestly and persistently in combatting the evils resulting from ignorance of physical laws, and there is no room for doubt that good success has attended the educational and reformative work" of his career.[9]

Orson's death churned a series of complex emotions in Charlotte and Lorenzo, both of whom had lost their own spouses by the time he passed. Samuel Wells had died in 1875, and Lydia Fowler had passed away in 1879. Still living in London with Lorenzo and the kids, Lydia worked as a physician and provided medical care to the poor. During one visit to a poverty-stricken neighborhood, she caught an illness and contracted pleuropneumonia, which, after a nine-week battle, brought her life to an end on January 26, 1879, at the age of fifty-six.

After Lydia, then Orson, the next to pass was Lorenzo. In 1893, he suffered a stroke that effectively paralyzed him and got him thinking about the country of his birth and his remaining sibling. His bond with Charlotte had always been special. So with the little life he had left, Lorenzo and his daughter moved to New Jersey to stay with Charlotte. But the two old siblings, both of whom were in their eighties, were able to reconnect for only a short time. A week after he arrived at Charlotte's house, Lorenzo passed away on September 2, 1896.

Despite her age, Charlotte remained active in phrenology for several more years, although usually in the form of correspondence from her New Jersey home. Still, newspapers took notice of her ongoing activity and remarkable longevity. One paper praised her as "the oldest woman publisher in the world." "Nearly her entire life," the paper summarized, "Mrs. Wells has been devoted to the advancement of the phrenological cause, contributing her intelligence, care, enthusiasm and rare activity to its extension in all parts of the civilized world." It was a peaceful end to her life when she passed away at her home on June 4, 1901.[10]

In Rosedale Cemetery, Orange, New Jersey, there is a Fowler family plot. Buried there are the Fowler parents, Horace and Martha; Lorenzo and his children (Lydia was buried in London); and Charlotte and Samuel Wells. Notably absent is Orson, who was buried unceremoniously in an unmarked grave in the Bronx. How or why he ended up there is a mystery, but apparently even his children felt disconnected from him at the time of his death. The distance of his grave from the graves of the other Fowlers now seems painfully fitting: The crumbling remnants of a once-great empire were scattered to different sites, never to be rebuilt.

WHAT HAPPENED TO the Empire of Skulls—or, more specifically, to the actual skulls that the Fowlers had amassed to build their empire? Like the siblings, they ended up as scattered, crumbling remnants.

Decades before Charlotte, Lorenzo, and Orson passed away, Nelson Sizer had spearheaded an effort to create the American Institute of Phrenology. Founded in 1866, it was designed to be an educational arm of the Fowler empire. Every year, a dozen or so students graduated from the Institute with a crisp certificate in practical phrenology, which supposedly authorized them to set up a private practice and extend the Fowler message. In 1884, when Charlotte was increasingly concerned about what would happen to the business after she passed, she gifted the Institute the Fowlers' massive collection of artifacts—skulls, busts, casts, paintings, and more. There were around two thousand items in the collection by that point, which, in one way, was a boon for the Institute; it now controlled one of the largest collections in the world, which Orson had started when he was still a student at Amherst College. But in another way, the gift was a burden, given that housing the collection and keeping it in good shape required ongoing maintenance, resources, and space.[11]

In 1892, Sizer launched a fundraising campaign to give the Institute and its collection a permanent home. He secured a decent contribution of $7,000 from an elderly couple long supportive of phrenology, and with that in hand he turned to the merchant Henry Sage for a contribution of $18,000. Sage declined, which led to a cascade of issues that ultimately ended Sizer's fundraising efforts. Practical phrenology didn't have the friends it once had. Without a permanent home, the artifacts of the Fowler collection were slowly dispersed and destroyed.[12]

Jessie Fowler (1856–1932), heiress of the crumbling Empire of Skulls

Although the siblings and skulls were scattered, there was one more Fowler doing her best to keep practical phrenology alive. In 1896, when Lorenzo returned to America to spend his final time on earth with his sister, he was accompanied by his daughter, Jessie Fowler. In her forties at the time, Jessie had grown up in England and had studied practical phrenology since birth. By the time she moved to the United States, she was, without question, the most active Fowler phrenologist on the planet, writing extensively for her father's publishing company in England and lecturing around the world, including in Australia. She came to America knowing full well that she was to take the reins of Fowler and Wells. Lorenzo was in no shape to do it, and both Charlotte and Nelson Sizer, who was still with the firm, were in the final years of their lives.

Jessie quickly discovered that running the firm was tough. To attract new readers, she had to publish an increasing number of works on paranormal, occult, and mystic beliefs, including reincarnation, ghosts, and the astral body. *Modern Ghost Stories*, by Emma

May Buckingham, was a Fowler and Wells publication that included chapters on "spectral illusions," the "spirit telegram," the "phantom child," and "the haunted chamber." Otherworldly spirits had long surrounded phrenology and the Fowlers, but the Spiritualist craze of the mid-nineteenth century was long past, and Jessie's turn to ghost stories was more about tapping into the occult market than about promoting a new scientific reality.

Another attempt at a new revenue stream came in the form of popular entertainment: Jessie read heads as a party trick. For only ten dollars, she showed up to "receptions or house parties" and provided "brief circle readings which would tell the leading points for each person and greatly interest all as a practical demonstration of Phrenology." To be sure, phrenology had always been partly a theatrical performance; the elder Fowlers were showmen after all. But at the dawn of the twentieth century, phrenology's primary value was as a parlor game, not much more.[13]

The year 1911 marked the passing of an era when the last issue of *The Phrenological Journal* was published. There was no fanfare or parting message; Jessie simply ran out of money and could no longer afford to publish it. She then sold the firm to a distant cousin, who basically ran it into the ground. Maintaining a small office in New York, Jessie continued to write and promote reform causes, but it was a relatively quiet end to her life when she passed away in October 1932.[14]

WHEN THE ESTEEMED European scientist Johann Spurzheim sailed for the United States in 1832, it was almost certain that phrenology would excite the nation. But no one could have predicted that it would actually be siblings from the backwoods of upstate New York who turned the science into a decades-long phenomenon. In a wonderfully democratic development, the Fowlers grabbed phrenology from its European progenitors and preached it to the American people as exactly what the nervous, energetic, adolescent nation needed to move forward. Practical phrenology was the right science at the right time, speaking to hopes and fears, problems and possibilities, realities and reforms, the present and the future. It propelled the era of reform, and the era of reform propelled it.

But not only that—the Fowlers were the right people at the right time because they effectively made phrenology go viral in the pre–Civil War decades. They understood the revelatory power of hands-on, tactile truth, of eloquent lectures and awe-inspiring demonstrations, of inexpensive, helpful literature, and of fascinating artifacts that people could take home for personal study. They built an empire through an entrepreneurial spirit, a shrewd business sense, an undaunted work ethic, and a carefully cultivated network of supporters who could spread their gospel message further and farther than they alone could. Of course, as with everything in life, their success required a good bit of serendipity, which they experienced. But Orson, Lorenzo, and Charlotte, along with their spouses and associates, knew how to attune their science to different people, different problems, and hope for the future.

After several decades of power and influence, the Empire of Skulls crumbled in familiar fashion. Scandals, deaths, financial downturns, waning public support, in-fighting, and a civil war—what the Fowlers built could not survive such tumult. Few empires can. But longevity was not the key to their influence; it was scope. Science, culture, religion, education, entertainment, the arts, business, health, homes, sex, race, justice, and more—the Fowlers shaped them all in just a few decades. In touching heads, they helped a nation grow up. In feeling skulls, they addressed the pressing needs of the era.

Today no one is a practicing phrenologist, except as a historical sideshow. Nevertheless, the Fowlers remain with us, albeit in some indirect, ephemeral ways. One such way is through the artifacts of their empire. The next time you're watching television and see the inside of a doctor's, therapist's, or professor's office, keep your eyes peeled for a phrenology bust. You may well see it, including the name on its base: "L. N. Fowler." In the past couple decades or so it has shown up on *Friends*, *House*, *Doctor Who*, *Crazy Ex-Girlfriend*,

Elementary, *Rick and Morty*, and other shows. You can even buy your own bust online, a relatively cheap porcelain replica. If you scan reviews of the product, you'll see people thrilled that they could buy it for decoration or as a gag gift for family and friends, often those in mental-health professions. You can also buy a variety of posters and wall art featuring imagery from the Fowler empire. Practical phrenology is with us today as shorthand for quirky science, whimsical decorations, and our shared, peculiar past.

To be sure, it is also with us in troubling ways, particularly in our racial legacy. The Fowlers were often at the leading, progressive edge of racial politics of the time, pushing for justice and equality and providing a scientific rationale for freedom. Yet they were still products of their era, and reading many of their statements today rightly gives us pause, especially in light of where their science went. It is possible to draw a line—not always straight, but a line nonetheless—from the work of phrenologists to the work of Herbert Spencer, champion of what is known as social Darwinism, which involved applying "survival of the fittest" to social, political, and economic contexts. These contexts often brought about racist, imperialist policies. From there we can draw a line to eugenics, or the attempt to increase supposedly "desirable" human traits by controlling who can reproduce and pass on their genes. Numerous countries and many U.S. states used the logic of eugenics to pass laws that permitted forced sterilization and treated human beings as livestock. In the U.S. South, eugenics was used to rationalize segregation and racial purity laws, which, in turn, inspired the Nazi regime to protect and promote a supposed master race across Europe.[15]

We need to take phrenology's falsehoods, errors, and harmful influences seriously. At the same time, we need to acknowledge that we are not too far away from the dynamics that undergirded the Empire of Skulls. Just as we can draw a line from the work of the Fowlers to horrors of the twentieth century, we can draw a line from their work to many intellectual, social, and cultural dynamics today.

Consider the movement known as the "quantified self." The idea is to use various devices—sleep trackers, smart watches, fitness bands, blood-oxygen monitors, and more—to render your mind and body as trackable numbers. From there, you can monitor those numbers to understand health patterns, where you can improve, where you are running into problems, and how you should take action. Supposedly these numbers give you better access to your mind and body—your true self—than simple reflection alone. Quantify yourself, the movement insists, and you can open up new pathways of progress. That's what the Fowlers offered Americans as well. They gave every sitter a chart marking the numerical value of his or her particular organs. Those numbers indicated where that person's mind was well developed, underdeveloped, and in need of cultivation or diversion. The numbers of the Fowler chart captured a true self, which was why it was often handed down to subsequent generations as a prized possession. That was certainly true for James Garfield, who, before ascending to the White House, received at least four phrenology exams, including by Lorenzo Fowler and Nelson Sizer. His charts are now preserved at the James A. Garfield National Historic Site in his hometown of Mentor, Ohio.[16]

The same kind of quantification shapes the at-home DNA-testing industry. If you spit in a tube and mail it off to a testing company, you will soon receive a report listing your genetic makeup, along with a health profile, description of your character, and ancestry map. You can even use your profile in match-making services to help you find a date. One online service, for instance, promises to forecast "romantic chemistry between people using DNA markers that play a role in human attraction" and via "personality types/psychology" that allow customers to "evaluate physical attraction based on your matches' photos." The offer is strikingly similar to promises of the Fowler empire, including those in Nelson Sizer's book *Right Selection in Wedlock* and Samuel Wells's book *Wedlock; or the Right Relation of the Sexes*. Americans of the nineteenth century could even send the siblings a photograph of themselves, along with five dollars, and receive an analysis of their character and advice on the kind of mate they should pursue.[17]

Personality tests, which have been popular for well over one hundred years, are part of this same impulse. You can characterize yourself scientifically using any number of tests—from Myers-Briggs or the Enneagram to big-five personality tests or DISC assessments. Not only are these tests freely available online, but they are given to children at school, to employees at big companies, even to parishioners at churches across the country. They render who you are on different charts, which, in turn, help you better know yourself and the people around you.

The tests offer us supposed scientific accuracy in response to our mysterious interiority, just as phrenology did. Every generation ponders what makes us tick, what motives us, what drives us, what compels us. And every generation comes up with its own ways to provide clarity amid the blooming, buzzing confusion of reality. The people of adolescent America relished their phrenological charts in much the same way we cherish our tracking devices, DNA tests, personality exams, and more. No doubt subsequent generations will concoct new, unforeseen ways of putting our mysterious interiority on a chart, and they will look askance at our numbers and charts in much the same way we look with a mixture of humor and suspicion at practical phrenology.

Ultimately, our generation, future generations, and the people of pre–Civil War America were looking for one thing—improvement. We all want to get better—to make ourselves better and to help others become better. Practical phrenology was an exciting way of doing that. It put the hope of progress in our hands, requiring only that we attend a lecture, receive an examination, or read a book. If that sounds familiar, it's the same message of the multibillion-dollar self-help industry. It's the message of Tony Robbins, Eckhart Tolle, and Deepak Chopra, among others. And it's the message that has captivated Americans for every generation since the Fowlers. Horatio Alger taught us to pull ourselves up by our bootstraps. Dale Carnegie taught us the right way to win friends and influence people. Norman Vincent Peale helped us realize our true potential through the power of positive thinking. Louise Hay inspired us to take control in her bestseller, *You Can Heal Your Life*. How helpful any one of these gurus, viewpoints, or programs is will, of course, vary wildly from person to person. The same was true of practical phrenology—an important forerunner to the self-help industry of today. For some people it was interesting but ultimately unhelpful. For others it was abject nonsense. And for still others, including Walt Whitman, it was central to their growth, development, and influence.

The hope of improvement for all, no matter where people started in life, was an invigorating, democratic message for a searching, democratic age. Having survived the nation's founding, Americans were peering anxiously to the future and hoping to find something that would set the country's bold political experiment on the right track. Into this restless dynamic stepped practical phrenology and its scientific proof that everyone could become a better version of themselves. Orson, Lorenzo, and Charlotte Fowler, along with their spouses and associates, energized a nation for science, improvement, and reform. They were wrong about a lot of things (as are big thinkers of every age), but they dedicated their lives to progress across American society. With a captivating new view of the body and mind, they crafted a gospel message, preached it far and wide, and built a memorable Empire of Skulls.

APPENDIX

DO-IT-YOURSELF PHRENOLOGY

TODAY PRACTICAL PHRENOLOGY IS A LOST SCIENCE AND art, but perhaps this book has got you interested in what hundreds of thousands, if not millions, of Americans experienced in the nineteenth century. Perhaps you're interested in learning what the Fowlers did in the examination room. Perhaps you'd like to give your family and friends a cranial exam. Here's how you can do that.

The Fowlers were adamant that the best way to learn phrenology was by practice. You ultimately need experience to examine heads effectively. But how to get started? You'll need some basic supplies, including a map of the phrenological organs and a chart you can fill out with numbers when giving an exam. Both of these documents appear at the end of this appendix.

When you are first starting out on your phrenological journey, it's best to learn the organs based on people you know and their character traits. Perhaps you have a stubborn neighbor. You can expect that his head will be well developed at Firmness (no. 14, involving decisiveness, stability, pursuit of purpose), which is at the very top of the cranium. Or perhaps your cousin is a drunk. You can feel her head for overly developed Alimentiveness (no. 8, involving the appetite for food and drink), which is right in front of the ears. Or perhaps your brother is a philanderer. You can feel his head at the bottom back region for robust Amativeness (no. 1, involving love and sexual attraction). You'll want to start with as many "known" characters as possible so you can experience what large or very large organs are like.

Whenever you examine a stranger, you'll need to start a bit more carefully. The first thing you should do is acquire a fabric tape measure, which will allow you to record the overall size of your sitter's head. To begin, place the tape measure at the middle of the forehead, right around Eventuality (no. 32, involving one's grasp of actions, events, and phenomena) or Comparison (no. 37, involving similarities and inductive reasoning), and wrap it around the most protruding point of the back of the head, which will likely be around Inhabitiveness (no. 4, involving love of home and patriotism) or Philoprogenitiveness (no. 2, involving love of kids and animals). The larger the measurement, the smarter your sitter tends to be (excepting conditions like hydrocephaly). For men, a large head is around twenty-three inches; for women, a large head is closer to twenty-two inches. The smaller the size from there, the less fully developed the sitter's brain.

After you have an overall measurement, it's time to start mapping the organs. There are two ways to begin the process. You can look at the head and see where it is the biggest—front, top, side, back, and so on. Start with that region and feel for the most prominent organ. Chances are that that organ will be large or very large, which means you can give it a 6 or 7 on the handy chart. Let's say the sides of the person's head protrude a bit. There's a good chance that your sitter will have large Cautiousness (no. 11, involving anxiety and the apprehension of danger) or Sublimity (letter B, involving grandeur and beholding what is magnificent). Get a good feel for that organ, not only so you have a sense of a large organ but because it will be a controlling factor in your sitter's character. Once you have a grasp on the big organ, branch out to other nearby organs, feeling their relative sizes as you go. The Fowlers recommended that you use the balls of your fingers as the point of contact with the head; that will give you the most accurate feel.

But the more exacting method for finding and measuring organs involves a little face geometry. Looking at a head in profile, start with the outer corner of the eye and picture a line extending to the top part of where the ear connects to the head (not the top of the ear itself, but the top point of connection). That is Destructiveness (no. 7, involving force and the drive to destroy and cause pain), which extends slightly above the top of the ear itself. Hold your hands horizontal at the sides of the sitter's head, with your fingers pointing toward the front of the head, and place your ring finger between the top of the ear and the head itself. Then spread your middle finger apart about an eighth of an inch. It will fall naturally on Secretiveness (no. 10, involving the tendency to conceal and hide). You will often feel a ridge between Destructiveness and Secretiveness, depending on which is smaller or larger. Note the sizes of those organs.

Then return to the imaginary line extending from the outer corner of the eye to Destructiveness, and picture it extending farther toward the back of the head about one inch. That's Combativeness (no. 6, involving resistance, opposition, and defense), and you can gauge its size by rotating your hand slightly back and down from its resting place on Destructiveness and Secretiveness, effectively dropping your elbow down. If you then move your index finger about one-eighth of an inch toward the back of the head, you'll feel Adhesiveness (no. 3, involving friendship and social connections). Another eighth of an inch back and slightly down the head is Philoprogenitiveness. In the bottom front of that is Conjugality (letter A, involving monogamy and union for life), and right below that is Amativeness. On the part of the head right behind the ear is Vitativeness (letter E, involving love of existence and fear of death). About three-quarters of an inch above Philoprogenitiveness is Inhabitiveness, and right above that is Concentrativeness (no. 5, involving unity of thought and feeling as well as continuity of action).

You've now covered the back-bottom portion of the head. Onto the side. Right in front of where the ear meets the head is Alimentiveness, and it usually extends about an inch forward. If you put your pinkie on Alimentiveness and spread your other fingers about three-quarters of an inch apart, your ring finger will fall on Acquisitiveness (no. 9, involving personal ownership and the drive to possess), immediately in front of which, by about three-quarters of an inch, is Constructiveness (no. 20, involving the ability to build, make, and invent). Put your ring finger back on Acquisitiveness and, with your fingers still spread, your middle finger will fall on the line between Sublimity and Cautiousness. Move that finger slightly forward and backward to gauge the size of those organs. Then, if you move the middle finger one-half inch forward of Sublimity, you can gauge Individuality (no. 24, involving the observation and individualization of things in the surrounding environment).

Now it's time for a little more face geometry. Look at the side of the head and imagine a straight vertical line extending up from the ear opening. At the very top of the head lies the organ of Firmness. One inch toward the back of the head is Self-Esteem (no. 13, involving self-respect and feelings of liberty and independence), and one-half inch below that, on an imaginary line heading back to the ear opening is Approbativeness (no. 12, involving regard for character and the approval of others). If you put your middle finger on that organ, with your hand aligned vertically, and if you spread your fingers apart by three-quarters of an inch, your ring finger will fall on Conscientiousness (no. 15, involving feelings of justice, accountability, and duty). Pivot your elbow forward so that your pinkie and ring fingers are on the same plane, and your pinkie will land on Hope (no. 16, involving the anticipation of future success and happiness).

You're now in the area of the head dedicated to spiritual, lofty affairs of the mind. Half an inch forward of Hope is Marvelousness (no. 17, involving wonder and belief in the supernatural), and right above that, on the top of the head, is Veneration (no. 18, involving worship and respect for sacred things). These organs connect us to God and the mystical side of life. Three-quarters of an inch or so forward of Veneration and Marvelousness are Benevolence (no. 19, involving kindness, sympathy, and care for others) and Imitation (no. 22, involving the ability to pattern and copy). Standing behind your sitter, with your hand positioned horizontally, place your index finger on Benevolence and spread your fingers about three-quarters of an inch apart. Your middle finger will fall on Imitation and your ring finger on Ideality (no. 21, involving imagination, taste, and fancy).

Now it's on to the front of the head and the intellectual faculties. As you look at the face straight on, focus on the bridge of the nose. Immediately above that, about half an inch, is Individuality; when it is large or very large, it will cause the eyebrows to arch downward. Immediately above that, and right below the center of the forehead, is Eventuality. Then right above that, at the center of the forehead, is Comparison. Above Comparison, and right where the head begins to slope back, is Human Nature (letter C, involving the understanding of other people, their character, and their motives). If you put your middle finger on Human Nature and spread your other fingers apart by three-quarters of an inch, your index finger will fall on Agreeableness (letter D, involving the ability to persuade others through charm and disposition). Half an inch below that are two organs that can be hard to distinguish, but practice makes perfect. One is Causality (no. 36, involving cause and effect and deductive reasoning), and the other is Mirthfulness (no. 23, involving the sense of what is funny, absurd, and ridiculous).

Returning to the bridge of the nose, a series of organs are tightly packed around the brow ridge. On the side of the bridge of the nose, heading in toward the eye, is Form (no. 25, involving shape and configuration). On top of that is Size (no. 26, involving an apprehension of magnitude, proportion, and bulk), and then lining the top of the eyebrow are Weight (no. 27, involving the apprehension of specific gravity, attraction, and force), Color (no. 28, involving the ability to see and understand color), Order (no. 29, involving physical arrangement and placement), and Calculation (no. 30, involving relationships of numbers and the ability to calculate). It can be difficult to locate each of these organs individually, but they often develop together according to a person's perceptive abilities. On top of the ridge of organs are Locality (no. 31, involving the relative position of things and places), Time (no. 33, involving the apprehension of the passage of time), and Tune (no. 34, involving melody, tone, and musicality). Again, they're tightly packed together, so only the most experienced phrenologists can find them. The organ of Language (no. 35, involving communication through words, sounds, signs, and symbols) is located behind the eye, and the best way to gauge it is visually. If a person's eyes protrude outward and downward, effectively bulging, the person's Language is large. If the eyes are sunken, Language is small.

Once you have recorded the size of each organ, it's time to translate those numbers into a character reading. This can take hundreds of heads to get right. But because you're probably not going to launch a career as a practical phrenologist, you should just stick to pointing out the organs that are particularly large or particularly small. Those are the ones that control the person's character, more or less. Large Amativeness, for instance, means a strong sex drive. Small Destructiveness is a sign of a meek, tender soul, unable either to cause or endure much suffering. Large Acquisitiveness means that the person is ceaselessly maximizing wealth in order to acquire more goods. Large Approbativeness is someone who constantly seeks the approval of others—a people pleaser. Small Firmness—or a dip on the top of the head—means that the person is impulsive and unstable. Large Marvelousness means the person lives by a kind of spiritual intuition and is able to commune with God and His creation. A person with small Benevolence is someone callous and uninterested in the well-being of others. Large Imitation is someone who fits in because she is able to follow patterns, especially behavior patterns set by others. Small Individuality is someone relatively oblivious to the world around him and only notices what other people point out. Large Order is someone who is systematic and methodical. Time is about the succession of events, so someone with a large organ is punctual and in touch with the rhythms of the day, while someone with a smaller version is always late and regularly forgets dates. A person with large Human Nature is a great judge of character and can accurately perceive the motives of others.

Every organ on your sitter's head has its size, and that size interacts with the size of every other organ. Properly gauging the network of variously sized organs across the head is a job for only the most experienced practical phrenologists. But don't worry; phrenology isn't real anyhow.

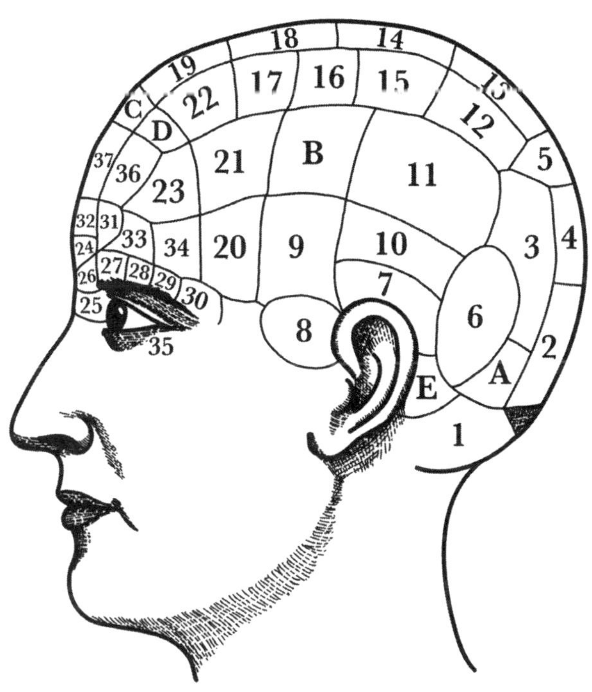

1. Amativeness
(A) Conjugality
2. Philoprogenitiveness
3. Friendship
4. Inhabitiveness
5. Concentrativeness
(E) Vitativeness
6. Combativeness
7. Destructiveness
8. Alimentiveness
9. Acquisitiveness
10. Secretiveness
11. Cautiousness
12. Approbativeness
13. Self-Esteem
14. Firmness
15. Conscientiousness
16. Hope
17. Marvelousness
18. Veneration
19. Benevolence
20. Constructiveness
21. Ideality
(B) Sublimity
22. Imitation
23. Mirthfulness
24. Individuality
25. Form
26. Size
27. Weight
28. Color
29. Order
30. Calculation
31. Locality
32. Eventuality
33. Time
34. Tune
35. Language
36. Causality
37. Comparison
(C) Human Nature
(D) Agreeableness

Phrenology Chart

Conditions	7 very large	6 large	5 full	4 average	3 moderate	2 small	1 very small
Size of Brain (in.)							
Domestic Propensities							
1. Amativeness....................							
(A) Conjugality..................							
2. Philoprogenitiveness.......							
3. Adhesiveness							
4. Inhabitiveness							
5. Concentrativeness							
Selfish Propensities							
(E) Vitativeness							
6. Combativeness							
7. Destructiveness							
8. Alimentiveness ....,....							
9. Acquisitiveness							
10. Secretiveness.................							
Selfish Sentiments							
11. Cautiousness							
12. Approbativeness...........							
13. Self-Esteem....................							
14. Firmness........................							
Moral Sentiments							
15. Conscientiousness.........							
16. Hope.............................							
17. Marvelousness							
18. Veneration.....................							
19. Benevolence...................							

Conditions	7 very large	6 large	5 full	4 average	3 moderate	2 small	1 very small
Semi-Intell. Sentiments							
20. Constructiveness							
21. Ideality							
(B) Sublimity							
22. Imitation							
23. Mirthfulness							
Observing Faculties							
24. Individuality							
25. Form							
26. Size							
27. Weight							
28. Color							
29. Order							
30. Calculation							
31. Locality							
Semi-Perceptive Faculties							
32. Eventuality							
33. Time							
34. Tune							
35. Language							
Reasoning Faculties							
36. Causality							
37. Comparison							
(C) Human Nature							
(D) Agreeableness							

ACKNOWLEDGMENTS

All writing is collaborative, and this book would not have been possible without the tremendous support of many people and institutions. The folks at Counterpoint Press have been incredible to work with from the very beginning. Special thanks to my editor, Dan Smetanka, for championing this project. Deirdre Mullane has been an ideal literary agent since I connected with her many years ago. A tireless worker on my behalf, she has shaped this book in innumerable ways.

I owe a special debt to the many research institutions that helped me secure the materials that made writing this book possible—Vanderbilt University Libraries, Cornell University Library Division of Rare and Manuscript Collections, Kislak Center for Special Collections at the University of Pennsylvania, Amherst College Archives and Special Collections, Francis A. Countway Library at Harvard University, New York Public Library Digital Collections, and the Internet Archive.

I have received indefatigable support from my friends and colleagues at Vanderbilt University, especially Cecelia Tichi, who connected me with Deirdre Mullane and immediately saw the promise of this book, and Bonnie Dow, who kindly listens to my strange ideas and makes a great cocktail. Huge thanks to my marvelous collaborators in the Program in Culture, Advocacy, and Leadership—Alex Jacobs, Mario Rewers, Victoria Hensley, Danyelle Valentine, Joe Bandy, Dana Nelson, Jesús Ruiz, and Kamaal Malak. We are building something significant. Special shout-out to my partner in crime and in beer brewing, Gabi Torres, whose friendship and jokes keep me sane (mostly). Let's brew again this weekend.

Over the years I have been fortunate to receive encouragement and mentorship from some inspiring scholars, particularly Steve Lucas, the late Jim Aune, and the late Marty Medhurst. Their influence appears in everything I write.

Last but certainly not least, I have been blessed with an incredible family. Lots of love to my parents, who have always supported me, and to my son, Elliott, who loves quirky Americana just as much as I do. Above all, thanks to my amazing wife, Sarah, who understood immediately why phrenology and the Fowler family would make for a fascinating read.

NOTES

PROLOGUE: SECRETS OF THE PIRATE'S HEAD

1. For an excellent book on the pirate and these murders, see Rich Cohen, *The Last Pirate of New York: A Ghost Ship, a Killer, and the Birth of a Gangster Nation* (New York: Random House, 2019).
2. For the pirate's own account of the murders, see *The Life, Trial, Confession and Execution of Albert W. Hicks* (New York: De Witt Publishing House, 1860), 62–66.
3. See "A Marine Mystery," *New York Tribune*, March 22, 1860; and "New York City News," *Brooklyn Eagle*, March 22, 1860.
4. "Execution of Hicks, the Pirate," *The New York Times*, July 14, 1860.
5. *The Life, Trial, Confession and Execution of Albert W. Hicks*, 40.
6. For all the details of the character reading, see *The Life, Trial, Confession and Execution of Albert W. Hicks*, 67–68.
7. The most detailed account of Gall's life and work with Spurzheim is Stanley Finger and Paul Eling, *Franz Joseph Gall: Naturalist of the Mind, Visionary of the Brain* (Oxford, UK: Oxford University Press, 2019).
8. For the classic treatment of the era of reform among historians, see Henry Steele Commager, *The Era of Reform, 1830–1860* (Princeton, NJ: Van Nostrand, 1960).
9. The following chapters will tell the story of the Fowlers in great detail. For another, excellent account, see Madeleine B. Stern, *Heads & Headlines: The Phrenological Fowlers* (Norman: University of Oklahoma Press, 1971).
10. O. S. Fowler, *Human Science* (New York: O. S. Fowler, 1870), 216.
11. Fowler, *Human Science*, 70.
12. For an overview of the Fowlers' work and the language of interiority, see Christopher Castiglia, *Interior States: Institutional Consciousness and the Inner Life of Democracy in the Antebellum United States* (Durham, NC: Duke University Press, 2008), ch. 5.
13. "Scientific Racism," Harvard University Library, n.d. guides.library.harvard .edu/scientificracism, accessed October 14, 2024; Martin Lund, "A Pre-history of Scientific Racism," MIT Press Reader, September 10, 2024,

thereader.mitpress.mit.edu/a-prehistory-of-scientific-racism/; "A History of Craniology in Race Science and Physical Anthropology," *Penn Museum*, n.d., www.penn.museum/sites/morton/craniology.php; Peter Leman, "How Profit and Prejudice Built a Family's Human Skull Collection," *Atlas Obscura*, October 2, 2019, www.atlasobscura.com/articles/fowler-wells-phrenology-cabinet; and Kim Severson, "Tom Colicchio Changes His Restaurant's Racially Tinged Name," *The New York Times*, August 22, 2017, www.nytimes.com/2017/08/22/dining/temple-court-fowler-and-wells-tom-colicchio.html.

14. O. S. Fowler, *Human Science* (privately printed, 1870), 219; and O. S. Fowler, *Hereditary Descent* (New York: O. S. & L. N. Fowler, 1843), 39.
15. O. Parker Jones, F. Alfaro-Almagro, and S. Jbabdi, "An Empirical, 21st Century Evaluation of Phrenology," *Cortex*, vol. 106 (September 2018): 26–35; and Oliver Wendell Holmes, "The Professor at the Breakfast Table," *Atlantic Monthly*, vol. 4 (August 1859): 241.
16. For a great overview of these themes, see Lee McIntyre, *The Scientific Attitude: Defending Science from Denial, Fraud, and Pseudoscience* (Cambridge, MA: MIT Press, 2019).

CHAPTER ONE: A MEETING OF MINDS

1. Harriet Beecher Stowe, *Men of Our Times* (Hartford, CT: Hartford Publishing, 1868), 529.
2. See "Sketches of Phrenological Biography: Henry Ward Beecher," *The Phrenological Journal*, vol. 92 (1891): 198–200. For a stellar biography of Beecher, see Debby Applegate, *The Most Famous Man in America: The Biography of Henry Ward Beecher* (New York: Image Books, 2006).
3. N. A. Shenstone, *Anecdotes of Henry Ward Beecher* (Chicago: J. C. Buckbee, 1887), 39–43.
4. See Applegate, *The Most Famous Man in America*, 96–97; and O. S. Fowler, *Synopsis of Phrenology* (New York: privately printed, 1837), 256.
5. Claude Moore Fuess, *Amherst: The Story of a New England College* (Boston: Little, Brown, and Company, 1935), 113. For additional references to their joint lectures and demonstrations, see Paxton Hibben, *Henry Ward Beecher: An American Portrait* (New York: George H. Doran, 1927), 58; and Lyman Abbott, ed., *Henry Ward Beecher, A Sketch of His Career* (New York: Funk & Wagnalls, 1883), 34.
6. Quoted in N. A. Shenstone, *Anecdotes of Henry Ward Beecher* (Chicago: J. C. Buckbee, 1887), 42.
7. Franz Joseph Gall, *On the Functions of the Brain and Each of Its Parts: With Observations on the Possibility of Determining the Instincts, Propensities, and Talents, of the Moral and Intellectual Dispositions of Men and Animals, by the Configuration of the Brain and Head*, vol. 5, trans. Winslow Lewis (Boston: Marsh, Cape, and Lyon, 1835), 8.
8. Stanley Finger and Paul Eling, *Franz Joseph Gall: Naturalist of the Mind, Visionary of the Brain* (Oxford, UK: Oxford University Press, 2019). See also the

various chapters in Paul Eling and Stanley Finger, eds., *Gall, Spurzheim, and the Phrenological Movement: Insights and Perspectives* (London, UK: Routledge, 2021).

9. For an excellent overview of the development of human understandings of the brain, see Stanley Finger, *Minds Behind the Brain: A History of the Pioneers and Their Discoveries* (Oxford, UK: Oxford University Press, 2000).
10. See Finger and Eling, *Franz Joseph Gall*, 57–59.
11. Finger and Eling, *Franz Joseph Gall*, ch. 3.
12. John van Wyhe, *Phrenology and the Origins of Victorian Scientific Naturalism* (London, UK: Routledge, 2004), 13–19.
13. Franz Joseph Gall, *On the Origin of the Moral Qualities and Intellectual Faculties of Man, and the Conditions of Their Manifestation*, vol. I (Boston: Marsh, Capen, and Lyon, 1835), 318.
14. For full details, see Finger, *Minds Behind the Brain*, ch. 3; and Finger and Eling, *Franz Joseph Gall*, ch. 4.
15. See Harry Whitaker, C. U. M. Smith, and Stanley Finger, eds., *Brain, Mind, and Medicine: Essays in Eighteenth Century Neuroscience* (New York: Springer, 2007), 13–28.
16. Finger, *Minds Behind the Brain*, 53.
17. See Timo Kaitaro, "Technological Metaphors and the Anatomy of Representations in Eighteenth-Century French Materialism and Dualist Mechanism," in *Brain, Mind, and Medicine*, ed. Harry Whitaker, C. U. M. Smith, and Stanley Finger, 335–44.
18. Quoted in Finger and Eling, *Franz Joseph Gall*, 44.
19. O. Parker Jones, F. Alfaro-Almagro, and S. Jbabdi, "An Empirical, 21st Century Evaluation of Phrenology," *Cortex*, vol. 106 (September 2018): 26–35.
20. Finger and Eling, *Franz Joseph Gall*, 43–47.
21. Finger and Eling, *Franz Joseph Gall*, 191–92.
22. In general, see Finger and Eling, *Franz Joseph Gall*, ch. 8.
23. For biographical details on Spurzheim, see John van Wyhe, "Johann Gaspar Spurzheim: The St. Paul of Phrenology," in *Gall, Spurzheim, and the Phrenological Movement*, eds. Paul Eling and Stanley Finger, 111–22.
24. For a general history of the end of the Holy Roman Empire, see Chip Wagar, *Double Emperor: The Life and Times of Francis of Austria* (Lanham, MD: Hamilton Books, 2018).
25. See Finger and Eling, *Franz Joseph Gall*, 318–20.
26. Finger and Eling, *Franz Joseph Gall*, ch. 9.
27. Quoted in Paul Eling and Stanley Finger, "Gall's German Enemies," in *Gall, Spurzheim, and the Phrenological Movement*, 67.
28. Finger and Eling, *Franz Joseph Gall*, 224.
29. Finger and Eling, *Franz Joseph Gall*, 233, 259.
30. Finger and Eling, *Franz Joseph Gall*, 234–35.
31. Finger and Eling, *Franz Joseph Gall*, 236.

32. See Paul Eling, Douwe Draaisma, and Matthijs Conradi, "Gall's Visit to the Netherlands," in *Gall, Spurzheim, and the Phrenological Movement*, 197–212.
33. For details on the reception of Gall and phrenology in France, see Marc Renneville, "Matter over Mind? The Rise and Fall of Phrenology in Nineteenth-Century France," in *Gall, Spurzheim, and the Phrenological Movement*, 173–84.
34. Finger and Eling, *Franz Joseph Gall*, 309.
35. See Harry Whitaker and Gonia Jarema, "The Split Between Gall and Spurzheim (1813–1818)," in *Gall, Spurzheim, and the Phrenological Movement*, 123–30.
36. See Stephen Tomlinson, *Head Masters: Phrenology, Secular Education, and Nineteenth-Century Social Thought* (Tuscaloosa: University of Alabama Press, 2005), 63.
37. For details on the origins of the name phrenology, see Régis Olry and Duane E. Haines, "Phrenology: Scheherazade of Etymology," in *Gall, Spurzheim, and the Phrenological Movement*, 7–14.
38. See van Wyhe, "Johann Gaspar Spurzheim," 117–20.
39. Quoted in Whitaker and Jarema, "The Split Between Gall and Spurzheim," 125.
40. For an excellent account of George Combe's influence on phrenology, see David Stack, *Queen Victoria's Skull: George Combe and the Mid-Victorian Mind* (London, UK: Hambledon Continuum, 2008).
41. See van Wyhe, *Phrenology and the Origins of Victorian Scientific Naturalism*, ch. 4.

CHAPTER TWO: COMING TO AMERICA

1. *Sailor's Magazine and Naval Journal*, vol. 4 (1831): 67.
2. Nahum Capen, *Reminiscences of Spurzheim and George Combe* (New York: Fowler and Wells, 1881), 6–7.
3. See John van Wyhe, "Johann Gaspar Spurzheim: The St. Paul of Phrenology," in *Gall, Spurzheim, and the Phrenological Movement: Insights and Perspectives*, eds. Paul Eling and Stanley Finger (London, UK: Routledge, 2021), 120.
4. Jon Meacham, *American Lion: Andrew Jackson in the White House* (New York: Random House, 2009), 49, 220.
5. Thomas Jefferson to John Adams, October 28, 1813, *Founders Online: National Archives*, founders.archives.gov/documents/Jefferson/03-06-02-0446. For an overview, see Lorri Glover, *Founders as Fathers: The Private Lives and Politics of the American Revolutionaries* (New Haven, CT: Yale University Press, 2014), ch. 5.
6. Meacham, *American Lion*, 57–64.
7. For an excellent overview of the social, cultural, and political world of "Jacksonian democracy," see the special forum *The Journal of the Early Republic*, vol. 39 (2019): 81–148.
8. *The Liberator*, January 1, 1831.
9. For a great book on religion in the nineteenth century, including the burned-over district, see Hadley Kruczek-Aaron, *Everyday Religion: An Archeology of Protestant Belief and Practice in the Nineteenth Century* (Gainesville: University of Florida Press, 2015).

10. Lyman Beecher, *Six Sermons on the Nature, Occasions, Signs, Evils, and Remedy of Intemperance* (New York: American Tract Society, 1827), 7–8.
11. See Todd Timmons, *Science and Technology in Nineteenth-Century America* (Westport, CT: Greenwood Press, 2005).
12. See Daniel Walker Howe, *What Hath God Wrought: The Transformation of America, 1815–1848* (Oxford, UK: Oxford University Press, 2007), 690–700.
13. See Howe, *What Hath God Wrought*, ch. 14. For more on the intersection of phrenology and democracy, see Arthur Wrobel, "Phrenology as Political Science," in *Pseudo Science and Society in Nineteenth-Century America*, ed. Arthur Wrobel (Lexington: University of Kentucky Press, 1987), 122–43.
14. For details on the Fowler family, including family trees, see "Fowler Family," box 4, untitled folder, Fowler and Wells Family Papers, Division of Rare and Manuscript Collections, Cornell University Library; and "New York Physicians," *The National Magazine*, vol. 19 (1894): 315.
15. O. S. Fowler, *Hereditary Descent* (New York: O. S. & L. N. Fowler, 1843), 61–66.
16. O. S. Fowler, *Hereditary Descent* (New York: Fowlers and Wells, 1847), 42.
17. O. S. Fowler, *Fowler's Works on Education and Self-Improvement* (New York: privately printed, 1844), 189; and Marion Sauerbier, "Horace Fowler's Family," *The Crooked Lake Review*, October 1988, www.crookedlakereview.com/articles/1_33/70ct1988/7sauerbier.html.
18. Fowler, *Fowler's Works on Education and Self-Improvement*, 182.
19. Fowler, *Fowler's Works on Education and Self-Improvement*, 208.
20. Fowler, *Hereditary Descent* (1843), 182–85; and O. S. Fowler, *Creative and Sexual Science* (Cincinnati, OH: Jones Brothers, 1870), 909–10.
21. Madeleine B. Stern, *Heads & Headlines: The Phrenological Fowlers* (Norman: University of Oklahoma Press, 1971), 7; and "Fowler Family," box 4, untitled folder, Fowler and Wells Family Papers, Division of Rare and Manuscript Collections, Cornell University Library.
22. Fowler, *Fowler's Works on Education and Self-Improvement*, 83.
23. O. S. Fowler, *Human Science* (privately printed, 1870), 572.
24. *The Philadelphia Inquirer*, August 16, 1821; and "Lectures on Phrenology in the Lunatic Asylum," *Cincinnati Gazette*, July 27, 1822.
25. See Frank R. Freemon, "Phrenology as Clinical Neuroscience: How American Academic Physicians in the 1820s and 1830s Used Phrenological Theory to Understand Neurological Symptoms," in *Gall, Spurzheim, and the Phrenological Movement: Insights and Perspectives*, eds. Paul Eling and Stanley Finger (London, UK: Routledge, 2021), 224–234.
26. See James Poskett, "*Django Unchained* and the Racist Science of Phrenology," *The Guardian*, February 5, 2013 www.theguardian.com/science/blog/2013/feb/05/django-unchained-racist-science-phrenology.
27. James Poskett, *Materials of the Mind: Phrenology, Race, and the Global History of Science, 1815–1920* (Chicago: University of Chicago Press, 2019), 119; and Charles Caldwell, *Elements of Phrenology* (Lexington, KY: Thomas T. Skillman, 1824), v, 2.

28. *The National Gazette*, March 18, 1822.
29. For an overview of Caldwell's early work on phrenology in the United States, see "Labours of Dr. Caldwell in Behalf of Phrenology," *American Phrenological Journal*, vol. 2 (1840): 451–56.
30. See *Aurora & Franklin Gazette*, May 19, 1825; Freemon, "Phrenology as Clinical Neuroscience," 227; and Fanny Trollope, *Domestic Manners of the Americans* (New York: Penguin, 1997), 55.
31. Capen, *Reminiscences of Spurzheim*, 6–8.
32. Charles E. Rosenberg, *The Cholera Years: the United States in 1832, 1849, and 1866* (Chicago: University of Chicago Press, 1987), 17; and John Noble Wilford, "How Epidemics Helped Shape the Modern Metropolis," *The New York Times*, April 15, 2008.
33. See Gerald Grob, *The Mad Among Us: A History of the Care of America's Mental Ill* (New York: Free Press, 1994); and Courtney E. Thompson, *An Organ of Murder: Crime, Violence, and Phrenology in Nineteenth-Century America* (New Brunswick, NJ: Rutgers University Press, 2021).
34. See Capen, *Reminiscences of Spurzheim*, 10; and Stephen Tomlinson, *Head Masters: Phrenology, Secular Education, and Nineteenth-Century Social Thought* (Tuscaloosa: University of Alabama Press, 2005), 225.
35. Capen, *Reminiscences of Spurzheim*, 10; and *Boston Morning Post*, August 30, 1832.
36. Capen, *Reminiscences of Spurzheim*, 22–24.
37. William Ellery Channing to Lucy Aikin, January 3, 1833, in *Correspondence of William Ellery Channing and Lucy Aikin*, ed. Anna Letitia le Breton (London, UK: Williams and Norgate, 1874), 159.
38. Capen, *Reminiscences of Spurzheim*, 26. See, for instance, "American Institute of Instruction," *Vermont Daily Chronicle*, September 14, 1832; *National Banner and Daily Advertiser*, October 16, 1832; and Capen, *Reminiscences of Spurzheim*, 28.
39. Capen, *Reminiscences of Spurzheim*, 30.
40. Quoted in "Dr. Spurzheim," *United States Telegraph*, November 15, 1832.
41. Quoted in Capen, *Reminiscences of Spurzheim*, 42.
42. Capen, *Reminiscences of Spurzheim*, 40.
43. See J. Collins Warren, "The Collection of the Boston Phrenological Society—A Retrospect," *Annals of Medical History*, vol. 3 (Spring 1921): 1–11; and Tomlinson, *Head Masters*, 226.

CHAPTER THREE: HEADS UP

1. For details of Lorenzo's life, see *Professor L. N. Fowler: A Sketch* (London: L. N. Fowler, n. d.), box 2, no folder, Fowler and Wells Family Papers, Division of Rare and Manuscript Collections, Cornell University Library.
2. See O. S. Fowler, *Human Science* (privately printed, 1870), 213.
3. Fowler, *Human Science*, 214.
4. Fowler, *Human Science*, 214–16.

5. O. S. Fowler and L. N. Fowler, *Phrenology Proved, Illustrated, and Applied* (New York: Edward Kearney, 1846), 258.
6. L. N. Fowler, *Lectures on Man* (New York: S. R. Wells, 1880), 28–29.
7. See no. 307, *Documents of the Assembly of the State of New-York, Fifty-Fourth Session* (Albany, NY: E. Croswell, 1831).
8. Fowlers, *Phrenology Proved*, 260.
9. Fowlers, *Phrenology Proved*, 260–61.
10. Fowlers, *Phrenology Proved*, 261.
11. Over the decades, the total number of organs fluctuated, even for the Fowlers. Sometimes they included new ones in their cartography, other times not. They also sometimes changed the name of the organs, depending on the audience they were trying to reach. For instance, they sometimes called Philoprogenitiveness "Parental Love" to make the meaning clear. The list of organs presented here follows the most common numbering and naming system in the Fowler corpus.
12. Fowlers, *Phrenology Proved*, 108–10.
13. Fowler, *Lectures on Man*, 171–72.
14. Fowler, *Lectures on Man*, 171–72.
15. See Charlotte Fowler Wells, "Sketches of Phrenological Biography," *Phrenological Journal and Science of Health*, vol. 93 (March 1892): 116–17; "Phrenology," *Daily Cincinnati Republican and Commercial Register*, May 6, 1835; "Phrenology," *Mississippi Free Trader and Natchez Gazette*, December 25, 1835; and "Communicated," *New Orleans Commercial Bulletin*, March 26, 1836.
16. Fowler, *Human Science*, 318.
17. Fowlers, *Phrenology Proved*, 305. Coverage appeared in several newspapers at the time, including in "Phrenology Tested," *United States Telegraph*, June 6, 1835.
18. Fowlers, *Phrenology Proved*, 306.
19. O. S. Fowler, *Phrenological Controversy* (Baltimore: John W. Woods, 1835), 5.
20. Fowler, *Phrenological Controversy*, 6–7.
21. Fowlers, *Phrenology Proved*, 354; and Fowler, *Phrenological Controversy*, 43.
22. O. S. Fowler, *Fowler's Works on Education and Self-Improvement* (New York: privately printed, 1844), 130.
23. O. S. Fowler and L. N. Fowler, *The Illustrated Self-Instructor in Phrenology and Physiology* (New York: Fowler and Wells, 1859), 83.
24. Fowler, *Fowler's Works on Education and Self-Improvement*, 4.
25. For a full account of Sewall's grave-robbing adventure, see Christopher Benedetto, "'A Most Daring and Sacrilegious Robbery': The Extraordinary Story of Body Snatching at Chebacco Parish in Ipswich, Massachusetts," *New England Ancestors*, vol. 6 (2005): 31–34, 45.
26. *Washington National Influencer*, November 13, 1835, 3. Also see *Washington Globe*, November 14, 1835, 3.
27. Fowlers, *Phrenology Proved*, 282–85.
28. Fowler, *Human Science*, 222–23.
29. Thomas Sewall, *An Examination of Phrenology; in Two Lectures, Delivered to*

the Students of the Columbian College, D. C., Feb. 1837 (Washington City: B. Homans, 1837), 59.

30. Quoted in Fowlers, *Phrenology Proved*, 371.
31. *New York Herald*, October 12, 1836.
32. Fowlers, *Phrenology Proved*, 324–26.

CHAPTER FOUR: A CABINET OF CURIOSITIES

1. See the typed, untitled speech transcript in box 3, folder 3, Fowler and Wells Family Papers, Division of Rare and Manuscript Collections, Cornell University Library.
2. "A Venerable Publisher," *Omaha Daily Bee*, August 9, 1896.
3. H. M. Slocum, "A Brief Sketch of a Woman Who Will Occupy a Place in History," *Daily Milwaukee News*, January 27, 1878.
4. *Philadelphia Botanic Sentinel*, February 15, 1838.
5. See "Phrenological and Physiognomy," *Philadelphia Inquirer*, March 31, 1840; and O. S. Fowler, *Human Science* (privately printed, 1870), 383.
6. Fowler, *Human Science*, 794–95.
7. Fowler, *Human Science*, 201. Also see Nelson Sizer, *How to Study Strangers by Temperament, Face, and Head* (New York: Fowler and Wells, 1895), 44.
8. Fowler, *Human Science*, 201.
9. See the advertisement for "Phrenological Rooms" in the *New York Morning Herald*, June 3, 1837.
10. "Old Clinton Hall," *American Phrenological Journal*, vol. 19 (1854): 72.
11. Untitled article, *Philadelphia Botanic Sentinel*, February 15, 1838, 200.
12. See "Phrenology," *Philadelphia Botanic Sentinel*, June 15, 1837, 339.
13. "Brevoort," *American Phrenological Journal*, vol. 5 (1843): 47; and "The Journal to Its Patrons," *American Phrenological Journal*, vol. 6 (1844): 5–6.
14. See "Brevoort," 47–48; and "Caution to the Public," *American Phrenological Journal*, vol. 5 (1843): 144.
15. O. S. Fowler, *Fowler's Works on Education and Self-Improvement* (New York: privately printed, 1844), 125–26.
16. Paterson Guardian, "A Place of Skulls," quoted in *American Phrenological Journal*, vol. 29 (1859): 11.
17. "Phrenology," *New York Tribune*, February 1, 1844.
18. *New York Herald*, December 5, 1844; and *New York Sunday Dispatch*, November 22, 1846.
19. See "Phrenology—Free," *New York Herald*, April 18, 1843; "Phrenology," *New York Herald*, January 13, 1844; and *New York Tribune*, May 5, 1846.
20. See "The Rapid Advancement of Phrenological Science," *American Phrenology Journal*, vol. 11 (1849): 166; and *Catalogue of Portraits, Busts, and Casts in the Cabinet of the American Institute of Phrenology* (New York: S. R. Wells, 1875).
21. For a comprehensive account of Cannon's life and crimes, see Michael Morgan, *Delmarva's Patty Cannon: The Devil on the Nanticoke* (Charleston, SC: The History Press, 2015).

22. "Patty Cannon—A Murderer," *The Phrenological Almanac for 1841* (New York: 135 Nassau Street, 1841), 37–38.
23. See O. S. Fowler and L. N. Fowler, *Phrenology Proved, Illustrated, and Applied* (New York: Printed for the Authors, 1837), 28; "Presentation of Animal Skulls, by Samuel P. Huey," *American Phrenological Journal,* vol. 11 (1849): 102; and Levi Reuben, "Phrenology: Its History and Doctrines, No. 2," *American Phrenological Journal,* vol. 34 (1861): 4.
24. See "Phrenological Developments and Character of James Eager, Executed for the Murder of Philip Williams, May 9, 1845," *American Phrenological Journal,* vol. 7 (1845): 263–68; and Fowler, *Human Science*, 681.
25. See *American Phrenological Journal*, vol. 10 (1848): 72; and "Chinese Skulls," *American Phrenological Journal*, vol. 10 (1848): 260.
26. See "Skulls of the White Bear and Walrus—Presented by Daniel Lee," *American Phrenological Journal*, vol. 10 (1850): 222; and "The Skull and Phrenological Character of a Sandwich Island Chief," *American Phrenological Journal*, vol. 18 (1853): 112.
27. Fowler, *Human Science*, 315.
28. Fowler, *Human Science*, 315–16.
29. See "Flat-Head Indian Skull," *American Phrenological Journal*, vol. 28 (1858): 48; and Fowler, *Human Science*, 1197.
30. See George Combe, *Notes on the United States of North America During a Phrenological Visit in 1838–9–40*, vol. I (Edinburgh: Maclachlan, Stewart, & Company, 1841), 180; and L. N. Fowler, *The Phrenological Developments and Characters of J. V. Stout, the Sculptor, and Fanny Elssler, the Actress* (New York: 135 Nassau Street, 1841), 36.
31. See "Cabinet of the Boston Phrenological Society," *American Phrenological Journal*, vol. 3 (1841): 189; and "Harrewaukay, the New Zealand Cannibal and Chief," *American Phrenological Journal*, vol. 7 (1845): 285.
32. See "Phrenological Developments of Dr. Samuel Thompson," *The Phrenological Almanac, and Physiological Guide, for the Year of Our Lord 1845* (New York: O. S. Fowler, 1845), 46; and *American Phrenological Journal*, vol. 7 (1845): 252.
33. See "Phrenological Societies," *American Phrenological Journal*, vol. 3 (1841): 188; and "Phrenology in America," *American Phrenological Journal*, vol. 8 (1846): 30.
34. See "A List of Specimens Designed for Phrenological Societies," *American Phrenological Journal*, vol. 10 (1848): 130; and "Phrenology in Canada," *American Phrenological Journal*, vol. 18 (1853): 67.
35. "The American Phrenological Society," *American Phrenological Journal*, vol. 11 (1849): 334.

CHAPTER FIVE: SCIENCE FOR THE PEOPLE

1. David Meredith Reese, *Humbugs of New York: Being a Remonstrance Against Popular Delusion; Whether in Science, Philosophy, or Religion* (New York: John S. Taylor, 1838), 110, 144.
2. Reese, *Humbugs of New York*, 63, 65, 74.

3. See Aileen Fyfe, *Steam-Powered Knowledge: William Chambers and the Business of Publishing, 1820–1860* (Chicago: University of Chicago Press, 2012), ch. 2.
4. Fyfe, *Steam-Powered Knowledge*, 38.
5. "Prospectus of the American Phrenological Journal and Miscellany," *Maryland Gazette*, August 23, 1838.
6. "The Past and Future of the Journal," *American Phrenological Journal*, vol. 4 (1843): 314–15.
7. "Introductory Statement," *American Phrenological Journal*, vol. 1 (1838): 4–12.
8. "The Past and Future of the Journal," 314–15. Also see "Fifty Years of Phrenology," *The Phrenological Journal and Life Illustrated*, vol. 80 (1885): 5–31.
9. "Miscellany," *American Phrenological Journal*, vol. 9 (1847): 40.
10. "The Journal to Its Patrons," *American Phrenological Journal*, vol. 6 (1844): 7–8.
11. "My Proposed Course," *American Phrenological Journal*, vol. 4 (1842): 1–8.
12. "Phrenological Developments of Four Casts Sent to the Editor for Phrenological Examination," *American Phrenological Journal*, vol. 8 (1846): 190.
13. "Phrenological Developments of Four Casts," 191.
14. "Phrenological Developments of Four Casts," 191.
15. "Examination, by the Editor, of Casts of Skulls, Numbered 1 and 2," *American Phrenological Journal*, vol. 8 (1846): 224–26.
16. "The Lives and Characters of the Casts Nos. 1 and 2, Whose Phrenological Developments Were Given at Page 223," *American Phrenological Journal*, vol. 8 (1846): 368–69.
17. "The Lives and Characters of the Casts Nos. 1 and 2," 371–72.
18. Nelson Sizer, *Forty Years in Phrenology* (New York: Fowler & Wells, 1888), 181.
19. Sizer, *Forty Years*, 181.
20. For these numbers, see "The Past and Future Course of the Journal," 316; "Miscellany," *American Phrenological Journal*, vol. 6 (1844): 168; "Friends of Our Cause," *American Phrenological Journal*, vol. 10 (1848): 40; "Valedictory," *American Phrenological Journal*, vol. 18 (1853): 132; "Valedictory," *American Phrenological Journal*, vol. 28 (1858): 81; and Theodore Peterson, *Magazines in the Twentieth Century* (Urbana: University of Illinois Press, 1956), ch. 1.
21. "Samuel R. Wells," *The Phrenological Journal and Life Illustrated*, vol. 60 (1875): 352.
22. Some of the best behind-the-scenes information about the daily operations of the publishing firm can be found in the letters of the Fowler family to their cousin, Paul Howe. See box 4, no folder, Fowler and Wells Family Papers, Division of Rare and Manuscript Collections, Cornell University Library.
23. "A Peep at the Phrenological Cabinet," *The Illustrated Phrenological Almanac for 1851* (New York: Fowlers and Wells, 1851), 33–34.
24. O. S. Fowler, *Hereditary Descent* (New York: O. S. & L. N. Fowler, 1843), 183.
25. See James C. Whorton, *Nature Cures: The History of Alternative Medicine in America* (Oxford University Press, 2002), ch. 4.
26. See Jane M. Adams, *Healing with Water: English Spas and the Water Cure, 1840–1960* (Manchester, UK: Manchester University Press, 2015); and Brett

Grainger, *Church in the Wild: Evangelicals in Antebellum America* (Cambridge, MA: Harvard University Press, 2019), 135.

27. See "The Water Cure, Bathing Included," *American Phrenological Journal*, vol. 7 (1845): 278. For the devastating toll that the boy's death took on Orson, see the family letters to Paul Howe, box 4, no folder, Fowler and Wells Family Papers, Division of Rare and Manuscript Collections, Cornell University Library.
28. "Water Cure in the City of New York," *Water-Cure Journal*, vol. 1 (1845): 16.
29. L. N. Fowler, "Phrenological Character of Dr. Joel Shew," *American Phrenological Journal*, vol. 11 (1849): 297–300.
30. "To the Friends of Hydrotherapy," *American Phrenological Journal*, vol. 11 (1849): 232; John B. Blake, "Mary Gove Nichols, Prophetess of Health," *Proceedings of the American Philosophical Society*, vol. 106 (1962): 219–34; and "Water-Cure Publications," *American Phrenological Journal* vol. 16 (1852): 96.
31. See, for example, the advertisement in *Merry's Museum and Parley's Playmate*, vol. 20 (July 1850): 33.
32. For an excellent account of the development of mesmerism and its spread to the United States, see Erika Janik, *Marketplace of the Marvelous: The Strange Origins of Modern Medicine* (Boston: Beacon Press, 2014), ch. 5.
33. See Eric T. Carlson, "Charles Poyen Brings Mesmerism to America," *Journal of the History of Medicine and Allied Sciences*, vol. 15 (1960): 121–32.
34. "Phrenology and Magnetism," *American Phrenological Journal*, vol. 4 (1843): 84.
35. "Phrenology and Animal Magnetism," *American Phrenological Journal*, vol. 4 (1843): 215. For a fuller telling of the story, see "Animal Magnetism," *American Phrenological Journal*, vol. 5 (1844): 69–70.
36. "Phrenology and Animal Magnetism," 216; and "Phreno Magnetism," *American Phrenological Journal*, vol. 5 (1844): 39–40.
37. "The Philosophy of Life," *American Phrenological Journal*, vol. 7 (1845): 2; and O. S. Fowler, *Human Science* (privately printed, 1870), 204–205.
38. See "City Items," *New York Tribune*, June 5, 1850. Among the best books on the Fox sisters and the Spiritualist movement is Barbara Weisberg's *Talking to the Dead: Kate and Maggie Fox and the Rise of Spiritualism* (New York: HarperOne, 2004).
39. Andrew Jackson Davis, *The Principles of Nature, Her Divine Revelations, and a Voice to Mankind* (Boston: Colby & Rich, 1847).
40. For Charlotte's handwritten account of these séances, see box 2, folder 1, Fowler and Wells Family Papers, Division of Rare and Manuscript Collections, Cornell University Library.
41. "Phrenological Description of Andrew J. Davis, the Poughkeepsie Clairvoyant," *American Phrenological Journal*, vol. 9 (1847): 351–53.
42. "Clairvoyance: Its Harmony with the Known Laws of Mind," *American Phrenological Journal*, vol. 10 (1848): 30; and "Reviews," *American Phrenological Journal*, vol. 13 (1851): 119.
43. S. B. Brittan and Dr. B. W. Richmond, *A Discussion of the Facts and Philosophy of Ancient and Modern Spiritualism* (New York: Partridge & Brittan, 1853), 13.

44. See "Literature and Logic of 'The Interior,'" *The National Magazine*, vol. 1 (1852): 355; "Spiritual Materialism," *Putnam's Monthly*, vol. 4 (1854): 158–72; and Hiram Mattison, *Spirit Rapping Unveiled! An Exposé of the Origin, History, Theology and Philosophy of Certain Alleged Communications from the Spirit World* (New York: Mason Brothers, 1853), 117–18; and Madeleine B. Stern, *Heads & Headlines; the Phrenological Fowlers* (Norman: University of Oklahoma Press, 1971), 151.

CHAPTER SIX: CELEBRITY SCIENCE

1. For contemporaneous coverage of Lind's concert, including these quotations, see "Jenny Lind's First Concert," *New York Tribune*, September 12, 1850; "Jenny Lind's First Concert in New York," *New York Herald*, September 12, 1850; and "First Concert of Jenny Lind in America," *Boston Herald*, September 12, 1850.
2. "Jenny Lind's First Concert in New York," *New York Herald*, September 12, 1850; and "The Jenny Lind Concert," *Brooklyn Daily Eagle*, September 12, 1850.
3. Lorenzo Fowler, *Thinkers, Authors, Speakers: A Lecture* (London: W. Tweedie, 1863), 13–14.
4. "Jenny Lind," *American Phrenological Journal*, vol. 14 (1851): 28–31.
5. See Bonnie Carr O'Neill, *Literary Celebrity and Public Life in the Nineteenth-Century United States* (Athens: University of Georgia Press, 2017).
6. For a great overview of Barnum's life, see Robert Wilson, *Barnum: An American Life* (New York: Simon and Schuster, 2019).
7. See Rachel E. Walker, *Beauty and the Brain: The Science of Human Nature in Early America* (Chicago, IL: University of Chicago Press, 2023), 48.
8. See Adrienne Saint-Pierre, "My Character and Peculiarities So Correctly Portrayed," The Barnum Museum, August 6, 2021, barnum-museum.org/my-character-and-peculiarities-so-correctly-portrayed/; and P. T. Barnum, *The Life of P. T. Barnum* (New York: Redfield, 1855), 366.
9. "Charles S. Stratton," *Phrenological Almanac*, vol. 10 (1848): 42–43.
10. "Miss Sylvia Hardy," *American Phrenological Journal*, vol. 21 (1855): 120.
11. "Phineas T. Barnum," *American Phrenological Journal*, vol. 16 (1852): 100–101.
12. For the lives of Chang and Eng, see Yunte Huang, *Inseparable: The Original Siamese Twins and Their Rendezvous with American History* (New York: Liveright, 2019); and Joseph Andrew Orser, *The Lives of Chang and Eng: Siam's Twins in Nineteenth-Century America* (Chapel Hill: University of North Carolina Press, 2015).
13. "The Phrenological Character of Chan and Eng, the Siamese Twins, with a Likeness," *American Phrenological Journal*, vol. 8 (1846): 316–17.
14. "Chang and Eng," *American Phrenological Journal*, vol. 19 (1854): 48.
15. Quoted in "Phrenological Character of Professor Alexander Campbell," *American Phrenological Journal*, vol. 9 (1847): 361–65; and "Horace Mann," *American Phrenological Journal*, vol. 25 (1857): 123.
16. See "Henry Clay," *American Phrenological Journal*, vol. 16 (1852): 36; quoted in "Phrenological Character of Professor Alexander Campbell," *American*

Phrenological Journal, vol. 9 (1847): 235; and "The Phrenology, Physiology, and Mentality of Professor Samuel B. F. Morse," *American Phrenological Journal*, vol. 10 (1848): 22.

17. "Phrenology in Springfield, Mass.," *American Phrenological Journal*, vol. 11 (1849): 71–72.
18. John Brown Sr.'s phrenological reading from 1847 is printed in full in James Redpath, *The Public Life of Captain John Brown* (London: Thickbroom & Stapleton, 1860), 42–43.
19. John Brown Jr. to S. R. Wells, quoted in *American Phrenological Journal*, vol. 12 (1850): 38–39.
20. See John Brown Sr. to John Brown Jr., April 12, 1850, in *The Life and Letters of John Brown, Liberator of Kansas, and Martyr of Virginia*, ed. F. B. Sanborn (Boston: Roberts Brothers, 1891), 74; Madeleine B. Stern, *Heads & Headlines: The Phrenological Fowlers* (Norman: University of Oklahoma Press, 1971), 151; and "Phrenology in Northern Ohio," *American Phrenological Journal*, vol. 12 (1850): 326.
21. L. N. Fowler, *The Illustrated Phrenological Almanac for 1851* (New York: Fowlers and Wells, 1851), 33; "Phrenology in the Country," *American Phrenological Journal*, vol. 19 (1854): 96; and "Phrenology in Ohio," *American Phrenological Journal*, vol. 19 (1854): 136.
22. Stern, *Heads & Headlines*, 173.
23. See "Phrenological Description of W. Whitman by L. N. Fowler, New York, July 16, 1849," vol. 148, Walt Whitman Papers, 1841–1940. Archives and Manuscripts, Duke University Libraries.
24. "Lectures," *Brooklyn Daily Eagle*, March 5, 1846.
25. "A Chance for Men of Bad Character," *Brooklyn Daily Eagle*, March 7, 1846.
26. "Something About Physiology and Phrenology," *Brooklyn Daily Eagle*, March 10, 1847.
27. Quoted in Arthur Wrobel, "Walt Whitman and the Fowler Brothers: Phrenology Finds a Bard," dissertation completed at the University of North Carolina at Chapel Hill (1968), 86.
28. Nathaniel Mackey, "Phrenological Whitman," *Conjunctions*, vol. 29 (1997): 231.
29. Wrobel, "*Walt Whitman and the Fowler Brothers*," 102.
30. Stern, *Heads & Headlines*, 112.
31. Wrobel, "*Walt Whitman and the Fowler Brothers*," 105.
32. "An English and an American Poet," *American Phrenological Journal*, vol. 22 (1855): 90–91.
33. Quoted in Wrobel, "*Walt Whitman and the Fowler Brothers*," 110.
34. Horace Traubel, *With Walt Whitman in Camden, Vol. I* (Boston: Small, Maynard & Co., 1905): 385.

CHAPTER SEVEN: WHAT'S SEX GOT TO DO WITH IT?

1. Marilynn Wood Hill, *Their Sisters' Keepers: Prostitution in New York City, 1830–1870* (Berkeley: University of California Press, 1993), 36–37.

2. Timothy J. Gilfoyle, *City of Eros: New York City, Prostitution, and the Commercialization of Sex, 1790–1920* (New York: Norton, 1992), 34–46, 132.
3. William W. Sanger, *The History of Prostitution: Its Extent, Causes, and Effects Throughout the World* (New York: Harper & Brothers, 1858), 18, 22.
4. O. S. Fowler and L. N. Fowler, *The Illustrated Self-Instructor in Phrenology and Physiology* (New York: Fowler and Wells, 1857), 51.
5. L. N. Fowler, *Marriage: Its History and Ceremonies* (New York: Fowler and Wells, 1846), 81–84.
6. Fowler, *Marriage*, 86.
7. Fowler, *Marriage*, 67.
8. Max Bachert, *Is Prof. O. S. Fowler the Foulest Man on Earth?* (New York: privately printed, 1881).
9. O. S. Fowler, *Amativeness: or Evils and Remedies of Excessive and Perverted Sexuality* (New York: Fowlers and Wells, 1848), v.
10. Fowler, *Amativeness*, 10.
11. Fowler, *Amativeness*, 14.
12. Fowler, *Amativeness*, 19.
13. Fowler, *Amativeness*, 45.
14. O. S. Fowler, *Maternity; of the Bearing and Nursing of Children* (New York: Fowlers and Wells, 1849), iii–iv, 43.
15. Fowler, *Maternity*, 95.
16. Fowler, *Maternity*, 61, 64.
17. Fowler, *Maternity*, 81, 134.
18. Bloomer's life and work has not received the kind of scholarly attention it deserves. As a result, the best overview of her contributions remains D. C. Bloomer, *Life and Writings of Amelia Bloomer* (Boston: Arena Publishing, 1895).
19. Carla Bittel, "Woman, Know Thyself: Producing and Using Phrenological Knowledge in 19th-Century America," *Centaurus*, vol. 55 (2013): 17; Carol Faulkner, *Lucretia Mott's Heresy: Abolition and Women's Rights in Nineteenth-Century America* (Philadelphia: University of Pennsylvania Press, 2011), 102; and "Phrenological Chart of Lucretia Mott," *American Phrenological Journal*, vol. 17 (1853): 75–77.
20. Dorothy Sterling, *Ahead of Her Time: Abby Kelley and the Politics of Antislavery* (New York: W. W. Norton, 1991), 204; and Madeleine B. Stern, "Margaret Fuller and the Phrenologist-Publishers," *Studies in the American Renaissance* (1980): 230.
21. See *The Elizabeth Cady Stanton-Susan B. Anthony Reader: Correspondence, Writings, Speeches*, ed. Ellen Carol DuBois (Boston: Northeastern University Press, 1992), 271–74; and Elizabeth Cady Stanton to Daniel Cady, January 12, 1853, in *Elizabeth Cady Stanton: As Revealed in Her Letters, Diary, and Reminiscences*, eds. Theodore Stanton and Harriot Stanton Blatch (New York: Harper and Brothers, 1922), 47.
22. Quoted in Ida Husted Harper, *The Life and Work of Susan B. Anthony, Vol. 1* (Indianapolis: Bowen-Merril, 1898), 85–86.

23. Quoted in "Spirit of the New York Press," *Brooklyn Daily Eagle*, September 3, 1853.
24. For coverage of the convention, the mob disruption, and newspaper coverage, see *History of Woman Suffrage, Vol. 1*, eds. Elizabeth Cady Stanton, Susan B. Anthony, and Matilda Joslyn Gage (New York: Fowler and Wells, 1881), 546–77.
25. "Miss Lydia Folger Fowler," *The Phrenological Journal and Life Illustrated*, vol. 68 (1879): 288–89, 291.
26. "Mr. and Mrs. Fowler," *American Phrenological Journal*, vol. 11 (1849): 385.
27. "Shall Anatomical Knowledge Be Prohibited to Woman?," *American Phrenological Journal*, vol. 10 (1848): 94. Also see "Female Physicians," *American Phrenological Journal*, vol. 10 (1848): 204–207.
28. Madeleine B. Stern, *Heads & Headlines: The Phrenological Fowlers* (Norman: University of Oklahoma Press, 1971), 157.
29. "Lectures," *New York Daily Herald*, March 31, 1855.
30. *History of Woman's Suffrage, Vol. 1*, 178; and "The Bloomers in the Field," *Vermont Journal*, May 20, 1853.
31. Stern, *Heads & Headlines*, 229–30.
32. *History of Woman's Suffrage, Vol. 1*, 45.
33. "Southern Women's Bureau," *New York Daily Herald*, June 10, 1870; and "Mrs. Charlotte Fowler Wells," *The New Northwest* (Portland, Oregon), August 28, 1879.
34. "Mrs. Charlotte Fowler Wells," *The New Northwest*, April 20, 1877.
35. "Our New York Letter," *Daily Milwaukee News*, January 27, 1878; "The Close of the Public Career of a Notable Woman," *Brooklyn Daily Eagle*, January 14, 1878; and "A Venerable Publisher," *Omaha Daily Bee*, August 9, 1896. Also printed in *The Boston Globe*, August 30, 1896.

CHAPTER EIGHT: LIFE IN THE OCTAGON

1. "To the Public," *American Phrenological Journal*, vol. 14 (1851): 93; and "Fowlers & Wells in Boston," *American Phrenological Journal*, vol. 14 (1851): 118.
2. "Lectures on Man," *The Boston Post*, November 21, 1851.
3. "Our Boston Branch," *American Phrenological Journal*, vol. 16 (1852): 93; and "Phrenological Rooms," *Boston Evening Transcript*, March 17, 1852.
4. "Our Philadelphia House," *American Phrenological Journal*, vol. 19 (1854): 45; and "Lectures in Philadelphia," *American Phrenological Journal*, vol. 19 (1854): 96.
5. "Phrenology in Philadelphia," 96.
6. "308 Broadway," *American Phrenological Journal*, vol. 19 (1854): 88.
7. O. S. Fowler, *A Home for All: or a New, Cheap, Convenient, and Superior Mode of Building* (New York: Fowlers and Wells, 1848).
8. Fowler, *A Home for All*, 6.
9. Fowler, *A Home for All*, 8–9.
10. See Amy G. Richter, *At Home in Nineteenth-Century America: A Documentary History* (New York: New York University Press, 2015), 1–2; and Sally McMurry, *Families and Farmhouses in Nineteenth-Century America: Vernacular Design and*

Social Change (Oxford: Oxford University Press, 1988). For a great overview, see Irene Cheng, *The Shape of Utopia: The Architecture of Radical Reform in Nineteenth-Century America* (Minneapolis: University of Minnesota Press, 2023).

11. For more on amateur architects and the emerging architectural profession, see W. Barksdale Maynard, *Architecture in the United States, 1800–1850* (New Haven: Yale University Press, 2002).
12. See Cheng, *The Shape of Utopia*, ch. 1.
13. W. Barksdale Maynard, *Architecture in the United States, 1800-1850* (New Haven: Yale University Press, 2002), 35–40.
14. Fowler, *A Home for All*, 38–41.
15. Fowler, *A Home for All*, 61.
16. Fowler, *A Home for All*, 63.
17. Fowler, *A Home for All*, 71.
18. Fowler, *A Home for All*, 72.
19. Fowler, *A Home for All*, 47.
20. O. S. Fowler and L. N. Fowler, *The Illustrated Self-Instructor in Phrenology and Physiology* (New York: Fowler and Wells, 1857), 60.
21. Fowlers, *Illustrated Self-Instructor*, 58.
22. Fowlers, *Illustrated Self-Instructor*, 54–55.
23. "Mrs. L. N. Fowler, "Phrenology for the Use of Children in Schools and Families," *American Phrenological Journal*, vol. 9 (1847), 163–65.
24. Mrs. L. N. Fowler, *Familiar Lessons on Physiology, Vol. 1* (New York: Fowlers and Wells, 1847), vi.
25. Mrs. L. N. Fowler, *Familiar Lessons on Phrenology, Vol. 2* (New York: Fowlers and Wells, 1847), 29–30.
26. Fowler, *Familiar Lessons on Phrenology, Vol. 2*, 33.
27. Fowler, *Familiar Lessons on Phrenology, Vol. 2*, 65.
28. Fowler, *Familiar Lessons on Phrenology, Vol. 2*, 99, 113.
29. Fowler, *Familiar Lessons on Phrenology, Vol. 2*, 209.
30. O. S. Fowler, *A Home for All, or the Gravel Wall and Octagon Mode of Building* (New York: Fowler and Wells, 1853).
31. Fowler, *A Home for All*, 52.
32. Fowler, *A Home for All*, 65, 127.
33. Fowler, *A Home for All*, 138.
34. "Reform in House-Building," *The Literary Union*, vol. 1 (March 1850): 125; "A Home for All," *Circular*, December 20, 1853; and "Octagon House," *Christian Parlor Book*, vol. 1 (1855): 29.
35. "Record of the Times," *The Country Gentleman*, vol. 5 (1855): 352; S. H. Mann, "Octagon Houses," *The Cultivator*, vol. 6 (February 1858): 57; and John Johnston, "The Editors Shanty," *Anglo-American Magazine*, vol. 4 (March 1854): 324.
36. Estimates for the number of octagon houses built in the wake of Orson's book range from 500 to upward of 3,000. The actual number is likely below 1,000. See Rebecca Lawin McCarley, "Orson S. Fowler and a Home for All: The

Octagon House in the Midwest," *Perspectives in Vernacular Architecture*, vol. 12 (2005): 49–63; and Ellen Puerzer, *The Octagon House Inventory* (Eight-Square Publishing, 2011).

37. See "The Octagon Plans of Settlement," *American Phrenological Journal*, vol. 21 (1855): 17.
38. "Octagon Settlement Company," *American Phrenological Company*, vol. 22 (1855): 41.
39. For a wonderful firsthand account of the problems with Clubb's expedition, see Miriam Davis Colt, *Went to Kansas; Being a Thrilling Account of an Ill-Fated Expedition to That Fairy Land and Its Sad Results* (Watertown: L. Ingalls & Co., 1862).
40. For more on the panic of 1857, see James L. Huston, *The Panic of 1857 and the Coming Civil War* (Baton Rouge: Louisiana State University Press, 1987).
41. P'ter O'Dactyl, "The Late Prof. Fowlers' Home," *Brooklyn Standard Union*, September 28, 1888.
42. "Fowler's Queer Castle," *The New York Sun*, August 15, 1897.

CHAPTER NINE: HEADING TOWARD FREEDOM

1. "The March of Mind," *New Hampshire Statesman*, May 14, 1831.
2. "The March of Mind," *New Hampshire Statesman*, May 14, 1831; and *The United States Gazette*, July 21, 1831.
3. *The Liberator*, August 12, 1831.
4. Charles Caldwell, *Elements of Phrenology* (Lexington, KY: A. G. Meriwether, 1827), 238–39, 245, 262. Combe quoted in James Poskett, *Materials of the Mind: Phrenology, Race, and the Global History of Science, 1815–1920* (Chicago, IL: University of Chicago Press, 2019), 119.
5. Paul Erickson, "The Anthropology of Charles Caldwell, M.D.," *Isis*, vol. 82 (1981): 252–56.
6. O. S. Fowler and L. N. Fowler, *The Illustrated Self-Instructor in Phrenology and Physiology* (New York: Fowler and Wells, 1857), 41.
7. O. S. Fowler, *Hereditary Descent* (New York: O. S. & L. N. Fowler, 1843), 12.
8. Fowler, *Hereditary Descent*, 57.
9. Fowler, *Hereditary Descent*, 22–23.
10. Fowler, *Hereditary Descent*, 34.
11. Fowler, *Hereditary Descent*, 260.
12. Fowler, *Hereditary Descent*, 259.
13. Fowler, *Hereditary Descent*, 260–61.
14. Fowler, *Hereditary Descent*, 265.
15. "Miscellaneous," *The Liberator*, September 21, 1838.
16. See *William Lloyd Garrison: The Story of His Life Told by His Children, Vol. II: 1835-1840* (New York: The Century Co., 1885), 119.
17. For example, see William Lloyd Garrison to Helen E. Garrison, July 23, 1848, in *The Letters of William Lloyd Garrison, Vol. III: No Union with Slave-Holders, 1841-1849*, ed. Walter M. Merrill (Cambridge: Belknap Press of Harvard

University Press, 1973), 570; "American Phrenological Journal," *The Liberator*, March 19, 1840; and "Lydia Maria Child," *The Liberator*, September 16, 1841.

18. "American Phrenological Journal," *The Liberator*, March 19, 1840; "Fowler's Almanac," *The Liberator*, June 17, 1847; "David M. Reese, M. D.," *The Liberator*, November 18, 1836; "Phrenological," *The Liberator*, August 3, 1837; "The Fair," *The Liberator*, December 27, 1838; and "Lydia Maria Child," *The Liberator*, September 16, 1841.
19. "The Poughkeepsie Slave Case," *The Liberator*, September 11, 1851.
20. "Letter from L. N. Fowler, the Eminent Phrenologist," *The Liberator*, August 11, 1865.
21. L. N. Fowler to William Lloyd Garrison, July 3, 1867, Digital Commonwealth ark.digitalcommonwealth.org/ark:/50959/6h441z846.
22. Catherine H. Birney, *The Grimké Sisters: Sarah and Angelina Grimké* (Boston: Lee and Shepard, 1885), 166.
23. For the history of the revolution, Steeve Coupeau, *The History of Haiti* (Westport, CT: Greenwood Press, 2008). For a recent interpretation of the actions of Eustache Belin, see Cynthia A. Bouton, "Eustache's 'Amazing Ruses': Loyalty and Liberty in Saint-Domingue During the Haitian Revolution," *Slavery & Abolition*, vol. 42 (2021): 502–21.
24. For the full story, see Poskett, *Materials of the Mind*, ch. 2.
25. See "Character of Eustache," *American Phrenological Journal*, vol. 2 (1840): 177–82.
26. For reference to his use of this chart, see Martin Delany, *The Condition, Elevation, Emigration, and Destiny of the Colored People of the United States* (1852), ch. 6. Also see "Dr. J. J. Gould Bias," Colored Conventions Project, coloredconventions.org/black-mobility/delegates/dr-j-j-gould-bias/. That web page suggests that Bias wrote his own treatise on phrenology, which was not the case. He simply used Orson's blank chart and corresponding pamphlet.
27. "Phrenology in Ohio," *American Phrenological Journal*, vol. 8 (1846): 194.
28. "Henry E. Lewis," *The North Star*, December 22, 1848.
29. "Dr. Brown's Lecture," *The Christian Recorder*, March 23, 1861; "Letter from Newport, Del.," *The Christian Recorder*, February 24, 1866; and "Lectures," *The Christian Recorder*, November 29, 1862.
30. "Phrenology," *The Colored American*, September 23, 1837; and "Dr. Smith," *The Colored American*, September 30, 1837.
31. For an overview of Delany's Black Nationalist thought, see Tommie Shelby, "Two Conceptions of Black Nationalism: Martin Delany on the Meaning of Black Political Solidarity," *Political Theory*, vol. 31 (2003): 664–92.
32. *The North Star*, July 7, 1848.
33. A recent, brilliant biography of Douglass' life is David W. Blight, *Frederick Douglass: Prophet of Freedom* (New York: Simon & Schuster, 2018).
34. "Prospectus of the *American Phrenological Journal*," *The North Star*, July 14, 1848.
35. Frederick Douglass, "Cuba and the United States," *Frederick Douglass' Paper*, September 4, 1851.

36. J. R. Johnson, "Uncle William's Pulpit," *Frederick Douglass' Paper*, November 12, 1852.
37. "Lectures on Phrenology," *Frederick Douglass's Paper*, March 4, 1853.
38. Frederick Douglass, *The Claims of the Negro, Ethnologically Considered: An Address Before the Literary Societies of Western Reserve College, at Commencement, July 12, 1854* (Rochester: Lee, Mann, and Co., 1854).
39. See John Stauffer, Zoe Trodd, Celeste-Marie Bernier, Henry Louis Gates, and Kenneth B Morris, *Picturing Frederick Douglass: An Illustrated Biography of the Nineteenth Century's Most Photographed American* (New York: Liveright, 2015).
40. *The Life and Times of Frederick Douglass Written by Himself* (Boston: DeWolfe, Fiske, and Co., 1892), 299–300.

CHAPTER TEN: DECLINE OF AN EMPIRE

1. O. S. Fowler, *Creative and Sexual Science* (Cincinnati, OH: Jones Brothers, 1870 & 1875), 36.
2. Fowler, *Creative and Sexual Science*, 906, 1055.
3. "Prof. Fowler's Lectures," *The Louisville Daily Courier*, April 4, 1857; and *Republican Banner* (Nashville, Tennessee), May 11, 1858.
4. "Private Lectures to Men," *St. Louis Globe-Democrat*, February 27, 1857; "Prof. Fowler's Private Lectures," *The Courier-Journal* (Louisville, Kentucky), April 10, 1857; and "M. L. Association," *Evansville Daily Journal*, March 19, 1857.
5. O. S. Fowler, *The Family: In Three Volumes* (New York: O. S. Fowler Publisher, 1859), frontispiece advertisement.
6. See "Passengers," *Charleston Mercury*, January 5, 1858; "Phrenology and Physiology," *American Advocate* (Kinston, North Carolina), December 24, 1857; "Lectures on Phrenology," *Montgomery Weekly Advertiser* (Montgomery, Alabama), January 20, 1858; "Lectures on Phrenology," *New Orleans Crescent*, February 22, 1858; *Mississippian*, April 20, 1858; and *Memphis Daily Appeal*, April 24, 1858.
7. For example, see "Lectures on Phrenology," *Halifax British Colonist*, August 10, 1858; "Lectures on Man!!" *The Gazette* (Montreal, Quebec, Canada), August 30, 1859; "Lectures on Phrenology," *The Cleveland Leader*, April 9, 1859; and "Lecture at the Gothic Church, E. D.," *Brooklyn Daily Eagle*, February 2, 1860.
8. "Phrenology and Physiology," *Manchester Courier*, October 13, 1860; "How to Read Character," *Liverpool Mercury*, September 1, 1863; and "Phrenology," *Sheffield and Rotherham Independent*, April 8, 1864.
9. *Professor L. N. Fowler: A Sketch* (London: L. N. Fowler, n.d.), 6.
10. For an excellent history of the tumultuous decade, see Eric H. Walther, *The Shattering of the Union: America in the 1850s* (Lanham, MD: SR Books, 2004).
11. For more, see David L. Lightner, *Slavery and the Commerce Power: How the Struggle Against the Interstate Slave Trade Led to the Civil War* (New Haven, CT: Yale University Press, 2006).

12. A good overview of what is known as "bleeding Kansas" can be found in Michael E. Woods, *Bleeding Kansas: Slavery, Sectionalism, and Civil War on the Missouri-Kansas Border* (New York: Routledge, 2017).
13. See Stephen Puleo, *The Caning: The Assault That Drove America to Civil War* (Yardley, PA: Westholme, 2012).
14. See Tony Horwitz, *Midnight Rising: John Brown and the Raid That Sparked the Civil War* (New York: Henry Holt, 2011).
15. For a great one-volume history of the Civil War, see James M. McPherson, *Battle Cry of Freedom: The Civil War Era* (New York: Oxford University Press, 1988).
16. A great overview of the casualties, complete with graphs and charts, is available at "Civil War Casualties: The Cost of War: Killed, Wounded, Captured, and Missing," American Battlefield Trust, September 15, 2023, www.battlefields.org /learn/articles/civil-war-casualties.
17. See Bob Zeller, *The Blue and Gray in Black and White: A History of Civil War Photography* (Westport, CT: Praeger, 2005).
18. For more on Brady, see Roy Meredith, *Mathew Brady's Portrait of an Era* (New York: Norton, 1982). For Holmes's reflections, see Oliver Wendell Holmes, "Doings of the Sunbeam," *The Atlantic*, July 1863.
19. "William Howland," *Phrenological Journal*, vol. 50 (1870): 192.
20. "A Word to Editors," *American Phrenological Journal*, vol. 32 (1861): 109.
21. "Maj.-Gen George McClellan," *American Phrenological Journal*, vol. 34 (1861): 49; "Gen. O. M. Mitchell," *American Phrenological Journal*, vol. 36 (1862): 1; "William Tillman," *American Phrenological Journal*, vol. 32 (1861): 61; "Ambrose Burnside," *American Phrenological Journal*, vol. 25 (1862): 49; and "Abraham Lincoln," *American Phrenological Journal*, vol. 40 (1864): 97–98.
22. "Woman Can Do It," *American Phrenological Journal*, vol. 32 (1861): 133; and "The Journal," *American Phrenological Journal*, vol. 36 (1862): 20.
23. For the fire, see *The Phrenological Journal and Life Illustrated*, vol. 56 (1873): 126. For the financial panic, see Christoph Nitschke, "Theory and History of Financial Crises: Explaining the Panic of 1873," *Journal of the Gilded Age and Progressive Era*, vol. 17 (2018): 221–40.
24. "Up Broadway!" *American Phrenological Journal*, vol. 60 (1875): 336.
25. "Our Loss," *American Phrenological Journal*, vol. 60 (1875): 407.
26. Box 1, folder 7, Fowler and Wells Family Papers, Division of Rare and Manuscript Collections, Cornell University Library.
27. "Our New York Letter," *Milwaukee Daily News*, January 27, 1878. Similar praise for Charlotte appeared in "The Close of the Public Career of a Notable Woman," *Brooklyn Daily Eagle*, January 14, 1878.
28. See *Amherst College: Biographical Record of the Graduates and Non-Graduates*, eds. Robert S. Fletcher and Malcolm O. Young (Amherst, MA: Trustees of Amherst College, 1939), 30.
29. O. S. Fowler, *Human Science* (New York: O. S. Fowler, 1870), 618.
30. Fowler, *Human Science*, 4.

31. O. S. Fowler, *Sexual Science* (Philadelphia, PA: National Publishing Company, 1870), xii.
32. Fowler, *Creative and Sexual Science*, 41, 1055–56.
33. While *Sexual Science* was printed by the National Publishing Company, based in Philadelphia, *Creative and Sexual Science* was printed by Jones Brothers & Co., based in Cincinnati.
34. Fowler, *Creative and Sexual Science*, ix, 1064.
35. Fowler, *Creative and Sexual Science*, v, viii, ix, 129, 137, 139, 146.
36. Fowler, *Creative and Sexual Science*, 486, 488, 489.
37. Fowler, *Creative and Sexual Science*, 517, 521, 578, 583.
38. Fowler, *Creative and Sexual Science*, 713–14, 716, 741.
39. Fowler, *Creative and Sexual Science*, 700, 703, 873–74.
40. Fowler, *Creative and Sexual Science*, 111.
41. Fowler, *Creative and Sexual Science*, 1064–65.
42. "Our New Departure," *The Phrenological Journal and Science of Health*, vol. 71 (1880): 220.
43. "A Phrenologist's Trouble," *The Boston Globe*, March 20, 1881; "Phrenology Gets a Bump," *The Bismarck Tribune*, March 25, 1881; and "Dr. O. S. Fowler," *Chattanooga Daily Times*, April 27, 1881.
44. Max Bachert, *Is Prof. O. S. Fowler the Foulest Man on Earth?* (New York: privately printed, 1881).
45. *Times-Picayune*, June 21, 1881; and *The Kansas City Times*, June 18, 1881.
46. "New York: Fowler, the Phrenologist, Likely to Fun Afoul of Anthony Comstock," *Chicago Tribune*, June 10, 1881; and "Phrenology," *Chicago Tribune*, June 16, 1881.
47. "New York: Fowler, the Phrenologist, Likely to Fun Afoul of Anthony Comstock," *Chicago Tribune*, June 10, 1881.
48. Quoted in Bachert, "Is Prof. O. S. Fowler the Foulest Man on Earth?"; and "O.S. Fowler," *Chicago Tribune*, June 27, 1881.
49. Bachert, "Is Prof. O. S. Fowler the Foulest Man on Earth?"

EPILOGUE: THE HOPE OF IMPROVEMENT

1. Charles Neider, ed., *The Autobiography of Mark Twain* (New York: Harper, 1959), 64–65.
2. Neider, *The Autobiography of Mark Twain*, 65–66.
3. See Burton J. Bledstein, *The Culture of Professionalism: The Middle Class and the Development of Higher Education in America* (New York: W. W. Norton and Company, 1978).
4. See Charles Darwin, *The Variation of Animals and Plants under Domestication*, 2nd ed. (London: John Murray, 1875), 83.
5. Erica Lilleleht, "'Assuming the Privilege' of Riding Divides: Abigail Fowler-Chumos, Practical Phrenology, and America's Gilded Age," *History of Psychology*, vol. 18 (2015): 414–32.

6. "Prof. O. S. Fowler," *Chicago Tribune*, August 24, 1887; and "Phrenological Memories," *Rochester Democrat and Chronicle*, August 27, 1887.
7. "Death of the Veteran Phrenologist," *Omaha Daily Bee*, August 26, 1887; "The Death of O. S. Fowler," *The Fort Worth Gazette*, September 2, 1887; and *The Weekly Register* (Point Pleasant, West Virginia), September 14, 1887.
8. "The Prophet of Phrenology," *The Montana Record-Hearld*, September 9, 1887. The article appeared in many other papers as well, including the *Kingston Whig-Standard*, September 1, 1887.
9. "Orson S. Fowler," *The Phrenological Journal and Life Illustrated*, vol. 82 (1887): 196–98; and "The Late Professor O. S. Fowler," *The Phrenological Magazine*, vol. 3 (1887): 445–46.
10. "A Venerable Publisher," *Omaha Daily Bee*, August 9, 1896; also printed in *The Boston Globe*, August 30, 1896. For additional late-in-life coverage, see the *Boston Evening Transcript*, August 14, 1897.
11. Madeleine B. Stern, *Heads & Headlines: The Phrenological Fowlers* (Norman: University of Oklahoma Press, 1971), 234–35.
12. Stern, *Heads & Headlines*, 236–37.
13. Emma May Buckingham, *Modern Ghost Stories: A Melody of Dreams, Impressions, and Spectral Illusions* (New York: Fowler and Wells, 1906); and advertisement, *The Phrenological Journal and Science of Health*, vol. 117 (1904): 7.
14. Stern, *Heads & Headliners*, 256–59; and "Jessie A. Fowler, Phrenologist, Dies," *The New York Times*, October 16, 1932.
15. See Daniel J. Kevles, *In the Name of Eugenics: Genetics and the Uses of Human Heredity* (Cambridge, MA: Harvard University Press, 1995).
16. See Mary Lintern, "Phrenology in Victorian America," James A. Garfield National Historic Site, U.S. National Park Service, www.nps.gov/articles/000/phrenology-in-victorian-america.htm, accessed June 3, 2025.
17. See the homepage of DNA Romance, www.dnaromance.com, accessed June 2, 2025; and *Illustrated Descriptive Catalogue on Phrenology* (New York: Fowler & Wells, n.d.), in box 1, folder 2, MS Collection 1418, University Archives and Records Center, University of Pennsylvania.

IMAGE CREDITS

12 Internet Archive
15 Internet Archive
16 Internet Archive
17 Fowler and Wells families papers, #97. Division of Rare and Manuscript Collections, Cornell University Library
28 Internet Archive
38 Internet Archive
45 Internet Archive
55 Fowler and Wells families papers, #97. Division of Rare and Manuscript Collections, Cornell University Library
59 Internet Archive
67 Warren Anatomical Museum Collection, Center for the History of Medicine in the Francis A. Countway Library, Harvard University
76 Internet Archive
92 Kislak Center for Special Collections, Rare Books and Manuscripts, University of Pennsylvania Libraries
106 Internet Archive
125 Internet Archive
129 Fowler and Wells families papers, #97. Division of Rare and Manuscript Collections, Cornell University Library
131 Fowler and Wells families papers, #97. Division of Rare and Manuscript Collections, Cornell University Library
139 Internet Archive
140 Internet Archive

153 Library of Congress, Washington, D.C.
167 Internet Archive
172 New York Public Library Digital Collections
175 Internet Archive
185 Fowler and Wells families papers, #97. Division of Rare and Manuscript Collections, Cornell University Library
191 Fowler and Wells families papers, #97. Division of Rare and Manuscript Collections, Cornell University Library
198 Internet Archive
199 Internet Archive
203 Internet Archive
213 Internet Archive
226 Fowler and Wells families papers, #97. Division of Rare and Manuscript Collections, Cornell University Library
233 Fowler and Wells families papers, #97. Division of Rare and Manuscript Collections, Cornell University Library
240 Internet Archive
243 Library of Congress, Washington, D.C.
258 Library of Congress, Washington, D.C.
261 Internet Archive
263 Fowler and Wells families papers, #97. Division of Rare and Manuscript Collections, Cornell University Library
276 Fowler and Wells families papers, #97. Division of Rare and Manuscript Collections, Cornell University Library
282 Fowler and Wells families papers, #97. Division of Rare and Manuscript Collections, Cornell University Library

© Sarah Stob

PAUL STOB is the director of the program in Culture, Advocacy, and Leadership at Vanderbilt University, where he also teaches in the Department of Communication Studies and the program in Communication of Science and Technology. He is the coauthor of *The Art of Public Speaking*. Find out more at paulstob.com.